International Conference on Renewable Energy and Environment (ICREE 2017)

IOP Conference Series: Earth and Environmental Science
Volume 127

Toronto, Canada
1 – 3 November 2017

ISBN: 978-1-5108-5922-7
ISSN: 1755-1307

Printed from e-media with permission by:

Curran Associates, Inc.
57 Morehouse Lane
Red Hook, NY 12571

Some format issues inherent in the e-media version may also appear in this print version.

Printed by Curran Associates, Inc. (2018)

For permission requests, please contact the Institute of Physics
at the address below.

Institute of Physics
Dirac House, Temple Back
Bristol BS1 6BE UK

Phone: 44 1 17 929 7481
Fax: 44 1 17 920 0979

techtracking@iop.org

Additional copies of this publication are available from:

Curran Associates, Inc.
57 Morehouse Lane
Red Hook, NY 12571 USA
Phone: 845-758-0400
Fax: 845-758-2633
Email: curran@proceedings.com
Web: www.proceedings.com

Table of contents

Volume 127

2017 International Conference on Renewable Energy and Environment (ICREE 2017)

1–3 November 2017, Toronto, Canada

Accepted papers received: 21 February 2018
Published online: 5 March 2018

Preface

Papers

Renewable energy and new energy development

PREFACE

It is our great pleasure to welcome you to 2017 International Conference on Renewable Energy and Environment (ICREE 2017) which will be held in Toronto, Canada during Nov. 1-3, 2017. ICREE2017 is dedicated to issues related to Applied Renewable Energy and Environment.

The major goal and feature of the conference is to bring academic scientists, engineers, industry researchers together to exchange and share their experiences and research results, and discuss the practical challenges encountered and the solutions adopted. Professors from Portugal, Canada and China are invited to deliver keynote speeches regarding latest information in their respective expertise areas. It will be a golden opportunity for the students, researchers and engineers to interact with the experts and specialists to get their advice or consultation on technical matters, sales and marketing strategies.

This proceedings present a selection from papers submitted to the conference from universities, research institutes and industries. All of the papers were subjected to peer-review by conference committee members and international reviewers. The papers selected depended on their quality and their relevancy to the conference. The volume tends to present to the readers the recent advances in the field of Renewable Energy and Environment and various related areas, such as Renewable energy and new energy development, Environmental quality monitoring and assessment, Power Electronics Technology and Economic Strategy, etc.

We would like to thank all the authors who have contributed to this volume and also to the organizing committee, reviewers, speakers, chairpersons, sponsors and all the conference participants for their support to ICREE 2017.

Prof. Paulo S. D. de Brito, Instituto Politécnico de Portalegre, Portugal
Dec, 28 2017

IOP Conf. Series: Earth and Environmental Science **127** (2018) 011001 doi:10.1088/1755-1315/127/1/011001

Statement of Peer Review

All papers published in this volume of IOP Conference Series: Earth and Environmental Science (EES) (ISSN: 1755-1315) have been peer reviewed through processes administered by the proceedings Editors. Reviews were conducted by expert referees to the professional and scientific standards expected of a proceedings journal published by IOP Publishing.

IOP Conf. Series: Earth and Environmental Science **127** (2018) 011002 doi:10.1088/1755-1315/127/1/011002

Conference Committee

Advisory Committee

Prof. Bharat Bhargava, Purdue University, USA

Conference Chair

Prof. Paulo S. D. de Brito, Instituto Politécnico de Portalegre, Portugal

Program Chair

Prof. Martin Pudlik, University of Applied Sciences of Bingen, Germany
Prof. Dr. Hossam Gaber, University of Ontario, Canada
Prof. Zhongjie Huan, Tshwane University of Technology, South Africa

Technical Committee

Ali Elkamel, University of Waterloo, Canada
Adel Sharaf, University Newbrunswick, Canada
Diego Espinoza, Universidad Autonoma de San Luis Potosi, Mexico
Mirmehrdad Mirsanjari, Malayer University, Iran
Bao-rong Fu, Liaoning University, China
Jing Ren, Liaoning University, China
Haliza Abd.Rahman, Universiti Putra Malaysia, Malaysia
Jinliang Xu, North China Electric Power University, China
Frede Blaabjerg, Aalborg University, Denmark
Gheorghe-Daniel Andreescu, Politehnica University of Timisoara, Romania
Asher Brenner, Ben-Gurion University of the Negev, Israel
Ali Cheknane, Amar Telidji University of Laghouat, Algeria
Qiang Zhou, China Agricultural University, China
Yongtae Ahn, Gyeongnam National University of Science and Technology, Korea
Mehdi Bagheri, Nazarbayev University, Kazakhstan
Fares. AlMomani, Qatar University, Doha
Wanyang Dai, Nanjing University, China
Shafiqur Rehman, King Fahd University of Petroleum and Minerals, Saudi Arabia
Lin Chen, Institute of Fluid Science, Tohoku University, Japan
Hassan Ali, The University of Newcastle, Australia, Singapore
Weifeng He, Nanjing University of Aeronautics and Astronautics, China
Sreekanth K. J., Energy and Building Research Centre (EBRC), Kuwait
Mohamed Abbas, Unit of Solar Equipments Development UDES/CDER, Algeria
Cenk Onan, Yildiz Technical University, Turkey
Raja Sekhar Dondapati, Lovely Professional University, India
Ümmühan Başaran Filik, Anadolu University, Turkey
Yang Yue, Juniper Networks, USA
Dinesh Khanduja, National Institute of Technology, India

IOP Conf. Series: Earth and Environmental Science **127** (2018) 011003 doi:10.1088/1755-1315/127/1/011003

Peer review statement

All papers published in this volume of *IOP Conference Series: Earth and Environmental Science* have been peer reviewed through processes administered by the proceedings Editors. Reviews were conducted by expert referees to the professional and scientific standards expected of a proceedings journal published by IOP Publishing.

IOP Conf. Series: Earth and Environmental Science **127** (2018) 012001 doi:10.1088/1755-1315/127/1/012001

Techno-economic study of a distributed hybrid renewable energy system supplying electrical power and heat for a rural house in China

Jindou Yuan[1], Jinliang Xu[1] and Yaodong Wang[2]

School of Energy Power and Mechanical Engineering, North China Electric Power University, Beijing, PR China
Sir Joseph Swan Centre for Energy Research, Newcastle University, Newcastle, United Kingdom

yuanjindou@126.com

Abstract. Energy saving and emission reduction have become targets for modern society due to the potential energy crisis and the threat of climate change. A distributed hybrid renewable energy system (HRES) consists of photovoltaic (PV) arrays, a wood-syngas combined heat and power generator (CHP) and back-up batteries is designed to power a typical semi-detached rural house in China which aims to meet the energy demand of a house and to reduce greenhouse gas emissions from the use of fossil fuels. Based on the annual load information of the house and the local meteorological data including solar radiation, air temperature, etc., a system model is set up using HOMER software and is used to simulate all practical configurations to carry out technical and economic evaluations. The performance of the whole HRES system and each component under different configurations are evaluated. The optimized configuration of the system is found

1. Introduction

The energy crisis and the climate change have become major concerns and are gaining significant attention in China and the world. Energy cost and environmental effect is making energy management, optimization and utilizing more renewables a crucial issue in building operations and maintenance. Actually, prior to the development of coal in the mid 19th century, nearly all energy used was renewable. Over the years, many new technologies have come out trying to shift the conventional low-efficiency fossil fuel-fired energy mode to other alternative and efficient options. In 1954, Bell laboratories first successfully brought the silicon solar module which provided the 6% conversion efficiency to the outdoor [1]. During the early 20th century, gasification had been an important and common technology. However, using woods to generate electricity, they must be gasified first and then cleaned (eliminate the particles and char) and cooled down (remove the moisture) the producer gas, i.e. syngas before being burned in the engine [2]. Technology like heat recovery is a method to improve the efficiency of the whole system which is quite often implied in the Rankine Cycle (RC) or Organic Rankine Cycle (ORC) [3]. In the early 21st century, the emergence of distributed generation (DG) technologies carved out the new way for optimizing the energy consumption configuration [4]. Moreover, hybrid renewable energy systems (HRES) which terms to the complementary incorporation of several different renewable energy generations have been proved to be more reliable and efficient than single power system [5]. It is acknowledged that some of these technologies are quite mature while others are still developing.

Based on the technologies mentioned above, a distributed hybrid renewable energy system (HRES) consists of photovoltaic (PV) arrays, a wood-syngas combined heat and power generator (CHP) and back-up batteries is designed for a rural house, which is located in a mountainous area in China where power grid is hard to reach..

2. Proposed Conformation

For mountainous area, grid construction is highly expensive. Therefore, distributed generation is a perfect approach to solve the problem. More importantly, some local sources such as solar radiation and wood biomass are available for building an ideal HRES. As shown in Fig. 1, a stand-alone power system needs to satisfy the thermal and electrical load for a household. Without the support of the grid, the system should be flexible enough to adjust the power output according to the load change. Accordingly, energy storage system is designed and implemented. If the generation is more than load requirement, it has to be stored in the energy storing device. If the load is high and the generation is less, the storage device has to discharge [6].

3. Cost estimation

HOMER (Hybrid Optimization Model for Electric Renewables) is a simulation software and is developed by National Renewable Energy Laboratory. HOMER first assesses the technical feasibility of the hybrid microgrid system to make sure it can meet the load demand. It can simulate the operation of the system for an entire year. Second, it estimates the total net present cost (NPC) of the system which terms to the life-cycle cost including initial set-up costs, component replacement costs, and the purchasing power costs from the grid, operation and maintenance costs. NPC is given by the following formula [7],

$$NPC = C_{tot}/CRF(i,T_p) \quad (1)$$

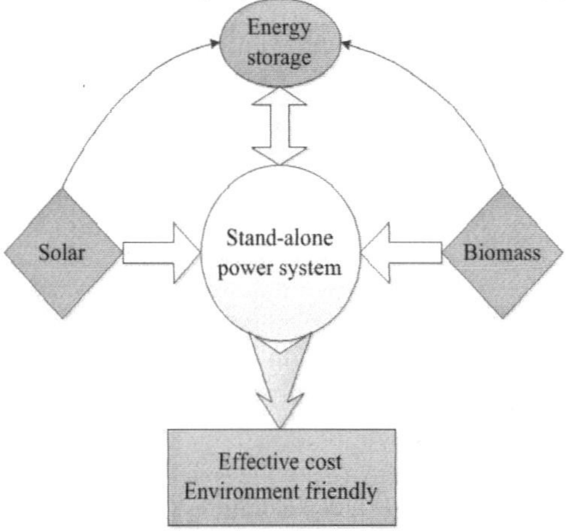

Figure 1. Flow chart of ideal HRES

where C_{tot} is the total annualize cost of the system (USD/year), i is the annual real interest rate (%),T_p is the project lifetime (10 years in this project), CRF(,) is the capital recovery factor, which is given by the following formula:

$$CRF(i,n) = i(1+i)^n/[(1+i)^n+1] \quad (2)$$

where n is the number of years.

In HOMER, the salvage costs (SC), which indicates the residual values of the components at the end of the project lifetime, are also taken into account in the estimation of the NPC, Homer gives the following formula:

$$SC = C_{RC}T_{rem}/T_{com} \quad (3)$$

where C_{RC} is the replacement cost of the component (USD), T_{rem} is the remaining life of the component (year), T_{com} is the lifetime of the component (year).

In Homer, the levelized cost of energy (COE) is given by:

$$COE = C_{tot}/E_{tot} \quad (4)$$

where E_{tot} is the total electricity consumption per year (kWh/year).

- 11 point Times or Times New Roman.
- The text should be set to single line spacing.
- Paragraphs should be justified.
- The first paragraph after a section or subsection heading should not be indented; subsequent paragraphs should be indented by 5 mm.

4. Cost study

4.1. Style and spacing

A typical rural house in Hubei province, south China (Latitude 30.7°N, Longitude 111.28°E) has been used as the case study to investigate and validate the efficacy of the proposed methodology. The load demand of the house including thermal and electric load is estimated by a four years average as shown in Fig. 2 and Fig. 3 respectively. It is understandable that the peak of the electric load appears in summer when the air condition starts to work while the peak of the thermal load is in winter. The specific data are gathered in Table I.

TABLE I SPECIFIC LOAD DEMAND

	Thermal load	**Electric load**
Average(kWh/d)	39.5	2.18
Average(kW)	1.65	0.0907
Peak(kW)	11.5	1.06
Load factor	0.143	0.0856

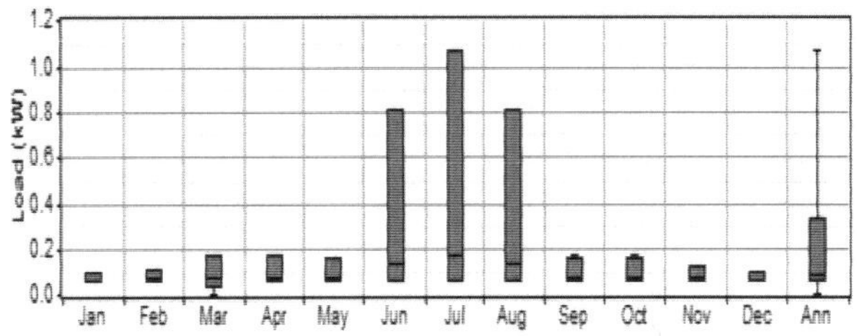

Figure 3. Monthly electric load data

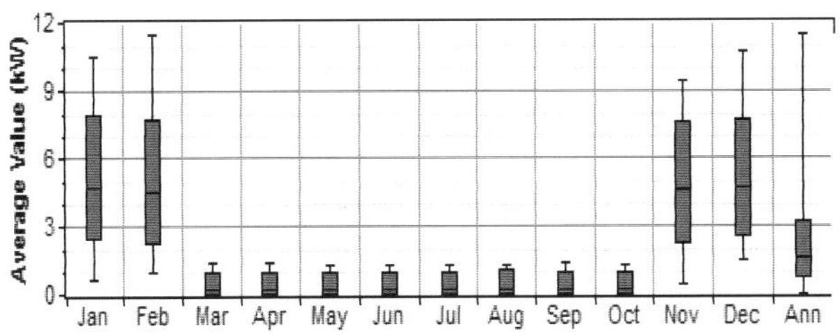

Figure 2. Monthly thermal load data

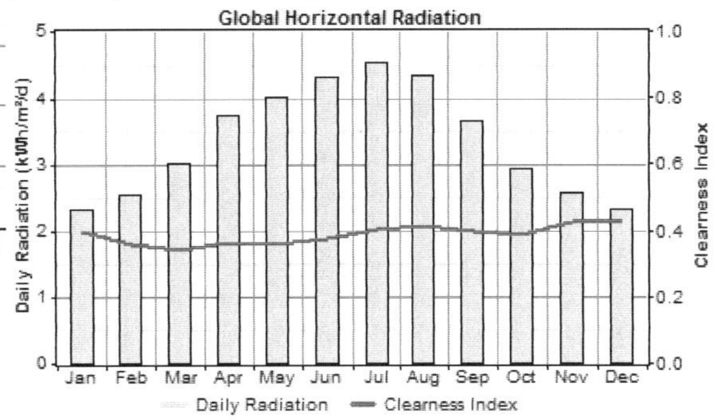

Figure 4. Average daily radiation and clearness index near the house

4.2. Sources Available

From the National Aeronautics and Space Administration (NASA) database on the RETscreen webpage, some basic information was acquired. Fig. 4 shows the daily solar radiation ranged from 2.31kWh/m^2 to 4.52kWh/m^2. Searching the mountains around the house, it turned out about 10 tons wood biomass available for use yearly (redundant branches and leaves in winter). Wood biomass is a necessary source for fueling the boiler (wood gasifier) and wood gas generator.

5. Model Simulation

The core concept of this study, in simple words, is to use the local sources to feed its own demand. As shown in Fig. 5, components consist of PV modules, T-105 battery which stand on the DC side and wood gasifier, wood gas generator which stand on the AC side. The converter plays a role of converting direct current or alternating current. To choose an appropriate battery for the energy storage system, the balance between generation and demand should be achieved. So,

$$\text{Required storage capacity} = \text{Max} \int \Delta P dt - \text{Min} \int \Delta P dt \quad (5)$$

$$\text{Number of batteries} \geq C_{Wh}/DOD \times \text{rated capacity of each capacity} \quad (6)$$

Where, ΔP is the magnitude difference between generated power (P_{gen}) and the demand (P_{dem}), C_{Wh} is the batteries total capacity, DOD is the depth of discharge [8].

The exhausted heat from the engine and the boiler is utilized to generate hot water and then supply to users. It can be achieved by installing some heat exchangers on the way of flowing exhausted wood gas. This can be considered as a typical small CHP system.

The control model in HOMER software includes load following (LF) model and cycle charging (CC) model [9].

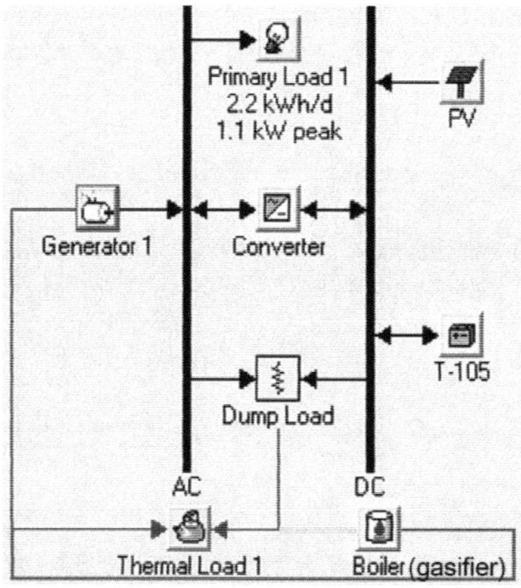

Figure 5. System structure simulation

6. Result and discussion

A total of 3780 (9×7×6×5×2) cases are made which the number is corresponding to the combinations of different sizes of components (generator × PV × converter × battery ×control model).

6.1. Mechanical Result

Many cases can satisfy the demand of the household. From the results of HOMER, it is found that the system of case 1 with 0.5kW of PV arrays, 0.65kW of wood gas generator, 3 unit batteries each of 1.35kWh and 0.5kW sized power converter in CC model obtains the best effectiveness, as shown in Table II and the details shown in Table III and table IV. Firstly, the lowest NPC and COE were achieved in this result (see in section B). Secondly, the whole system run perfectly to meet the demand of the household, which the unmet thermal load is 0 and the unmet electric load is small enough to be ignored. Thirdly, the excess thermal and electricity are the lowest

6.2. Economic Result

On the economic side, attentions often be paid on the COE and NPC. These two factors calculated automatically by the HOMER represent the economic feasibility of the whole system. The economic optimization result is shown in Table II.The lowest NPC and COE are 3572 USD and 0.351 USD/kWh respectively. It should be noted that Chinese government is promoting the development of distributed renewable energy [10] and matched preferential policies have been introduced including levy-refund and discount taxable income. So the real cost is much lower.

6.3. Emission Result
For the economic-optimal system shown in Table II, some important indicators of pollutant can be acquired by simulation which is shown in Table V. However, the biomass like wood could be considered renewable in a life cycle because the CO2 emitted by burning woods this year can be partly absorbed by the growing of new boughs next year through photosynthesis [11].

6.4. Sensitivity analysis
If we enhance the proportion of the PV arrays, for example, choosing the system of case 2 with 0.75kW PV and 0.65kW generator which electricity generated by PV takes 55% of the whole system, the emissions of the pollutant will have a slight decrease as shown in Table VI.

Table II Optimized results for the proposed configurations

Options	PV (kW)	Generator (kW)	battery	Converter (kW)	PV (kWh/yr)	Generator (kWh/yr)	Initial capital (USD)	Operating Cost(USD/yr)	NPC (USD)	COE (USD/kWh)
Case 1	0.5	0.65	3	0.5	518	618	1856	134	3572	0.351
Case 2	0.75	0.65	3	0.5	777	587	2292	134	4000	0.393
Case 3	0.25	1	3	0.5	259	755	2072	176	4320	0.425

Table III Thermal demand and supply

Thermal production(kWh/yr)	16155
Boiler output fraction (%)	83
Generator output fraction (%)	16
Excess thermal (kWh/yr)	1738
Unmet thermal load(kWh/yr)	0
Wood consumption(kg/yr)	3852

Table IV Electric demand and supply

Electric production(kWh/yr)	1136
PV output fraction (%)	46
Generator output fracion (%)	54
Excess electricity(kWh/yr)	190
Unmet electric load(kWh/yr)	3.13×10-7
Capacity shortage(kWh/yr)	0.568

Table V Emissions of Economic Optimization System

Pollutant	Emissions (kg/yr)
Carbon dioxide (CO2)	6490
Carbon monoxide (CO)	4.23
Particulate matter (PM)	0.427
Sulfur dioxide (SO2)	3.83
Nitrogen oxides (NOX)	1.39

TableVI Emissions of PV-enhance System

Pollutant	Emissions (kg/yr)
Carbon dioxide (CO2)	6407
Carbon monoxide (CO)	4.10
Particulate matter (PM)	0.414
Sulfur dioxide (SO2)	3.78
Nitrogen oxides (NOX)	1.34

Emission of carbon dioxide drops about 1.3%, carbon monoxide about 3.1%, particulate matter about 3.2%, sulfur dioxide about 1.3%, nitrogen oxides about 3.7%. However, the COE of this PV-enhanced system has a 12.0% increase. In fact, lower emissions always mean higher costs. Therefore, it is important to strike a balance between the profit making and environment protection.

7. Conclusion
In order to meet the target of promoting renewable energy in remote rural areas, this investigation of the feasibility of building a distributed HRES for a rural house was conducted in mountainous area of south China.

To find the optimized configuration, HOMER software was utilized to simulate all viable combinations in two different modes. The results had demonstrated that PV/wood gas generator/battery system with CC mode was evaluated to possess the lowest COE (0.351USD/kWh) and NPC (3572USD) for SAPS.

As for emissions, though the CO2 reaches 6490 kg/yr in the economic optimization system, the net emission can be thought as zero in a life cycle according to the properties of renewable biomass energy.

8. References

[1] Bajpai, Prabodh, Vaishalee Dash. "Hybrid renewable energy systems for power generation in stand-alone applications: A review," Renewable and Sustainable Energy Review, vol. 16, Jun. 2012, pp. 2926-2939.

[2] "Wood gas as engine fuel". Publications Division, Food and Agriculture Organization of the United Nations, Via delle Terme di Caracalla, 00100 Rome, Italy,1986.

[3] Shekh Nisar Hossain , Saiful Bari. "Waste heat recovery from the exhaust of a diesel generator using Rankine Cycle," Energy Conversion and Management, vol. 75, Nov. 2013, pp. 141–151.

[4] "The potential benefits of distributed generation and rate-related issues that may impede their expansion," US Department of Energy,2007

[5] Yaodong Wang, et al. "Modelling and simulation of a distributed power generation system with energy storage to meet dynamic household electricity demand," Applied Thermal Engineering, vol. 50, Jan. 2013, pp. 389-403.

[6] Anurag Chauhan, R.P. Saini. "A review on integrated renewable energy system based power generation for stand-alone applications: Configurations, storage options, sizing methodologies and control," Renewable and Sustainable Energy Reviews, vol. 38, Oct. 2014, pp. 99-120.

[7] Chong Li, Xinfeng Ge, Yuan Zheng, Chang Xu, Yan Ren, Chenguang Song, Chunxia Yang. "Techno-economic feasibility study of autonomous hybrid wind/PV/battery power system for a household in Urumqi, China," Energy, vol. 55, Jun. 2013, pp. 1-10..

[8] Leong Kit Gan, Jonathan K.H.Shek, Markus A.Mueller. "Hybrid wind-photovoltaic-diesel-battery system sizing tool development using empirical approach, life-cycle cost and performance analysis: A case study in Scotland," Energy Conversion and Management, vol.196, Dec. 2015, pp. 479-494.

[9] Saheb-Koussa, Djohra, Mustapha Koussa, and Nourredine Said. "Prospect of Wind-Diesel Generator-Battery Hybrid Power System: A Feasibility Study in Algeria," Journal of Wind Energy, vol. 2013, Jul. 2013.

[10] "The report for distributed energy in China," China Huadian Corporation, 2015

[11] See M Le Page, "The Great Carbon Scam," New Scientist, vol. 231, Sep. 2016, pp. 20–21

Acknowledgments

This research was supported by the Dr. Wang of Sir Joseph Swan Centre for Energy Research of Newcastle University and Prof. Xu of the School of Energy Power and Mechanical Engineering of North China Electric Power University.

IOP Conf. Series: Earth and Environmental Science **127** (2018) 012002 doi:10.1088/1755-1315/127/1/012002

Evaluating predictive models for solar energy growth in the US states and identifying the key drivers

Joheen Chakraborty[1,2] and Sugata Banerji[1]

[1] Department of Mathematics and Computer Science, Lake Forest College, 555 North Sheridan Road, Lake Forest, IL 60045, USA

[2] Worked as a summer research student

Correspondence should be directed to banerji@lakeforest.edu

Abstract. Driven by a desire to control climate change and reduce the dependence on fossil fuels, governments around the world are increasing the adoption of renewable energy sources. However, among the US states, we observe a wide disparity in renewable penetration. In this study, we have identified and cleaned over a dozen datasets representing solar energy penetration in each US state, and the potentially relevant socioeconomic and other factors that may be driving the growth in solar. We have applied a number of predictive modeling approaches - including machine learning and regression - on these datasets over a 17-year period and evaluated the relative performance of the models. Our goals were: (1) identify the most important factors that are driving the growth in solar, (2) choose the most effective predictive modeling technique for solar growth, and (3) develop a model for predicting next year's solar growth using this year's data. We obtained very promising results with random forests (about 90% efficacy) and varying degrees of success with support vector machines and regression techniques (linear, polynomial, ridge). We also identified states with solar growth slower than expected and representing a potential for stronger growth in future.

1. Introduction and previous work

The total installed solar capacity in the world grew by about 200-fold between 2000 (1.4 GW) and 2015 (257 GW). About 13.3% of global solar capacity was in the US as of 2016, but the growth of solar has been uneven. We have collected and cleaned extensive data on solar installations and their potential drivers over the past 17 years across all 50 US states, and applied a variety of machine learning and other modeling techniques in an attempt to construct the most effective model to predict solar growth as well as identify the relative importance of different factors that are driving it.

Most of the work to date that deals with predicting the growth in solar energy have been conducted by government and industry organizations. For example, [1] uses a model named National Energy Modeling System that aims to capture interactions of economic factors with energy supply, demand and prices. The work [2] provides an analysis of the solar growth in the US and [3] analyzes trends in global renewable energy installations. However, none of these works addresses state-level solar penetration or tries to identify the relative importance of contributing factors.

2. Our approach

For this research, we first created a list of candidate factors / independent variables (Table 1) that have the potential to drive the growth in solar energy in US states. There are two dependent variables that we used in our study to represent the solar growth: (1) % growth in solar penetration this year compared to the last year and (2) % growth in solar penetration next year compared to this year. We used the datasets listed in Section 3 to obtain data for the years 1997-2014.

After cleaning and consolidating the raw data, we systematically applied different modeling techniques including machine learning and regression to predict the two dependent variables using the independent variables. We also analyzed the model residuals to characterize state and year-specific bias in solar growth with respect to the model-predicted values.

3. Datasets

State Energy Data System (SEDS), U.S. Energy Information Administration [4]: SEDS provides comprehensive energy statistics at the state level including energy production and consumption by energy source. We extracted total energy production data from all sources per state for the years 1997-2014 as well as total solar energy (solar photovoltaic and solar thermal) consumed for the same period. We computed solar penetration per state from these data.

US Census Bureau, Median Household Income by State [5]: From this data source we obtained median household income per state for every year between 1997 and 2014 in terms of 2015 dollars.

Kaiser Family Foundation, Population distribution by race/ethnicity [6]: We obtained the demographic makeup of each state's population in 2015 broken down by White, Black, Hispanic and Asian. Given that these distributions generally change slowly over time and lack of easily available and reliable historical data, we assumed the breakdown to remain unchanged over the study period.

National Renewable Energy Laboratory (NREL) Sun Index [7, 8]: This per state index captures the solar energy potential of the state based on the total sunlight received after factoring in daylight duration, cloud cover, etc. Since solar potential of a region changes extremely slowly, we used the same values of the index for the entire study period although the sun index available was for 2006.

U.S. Census Bureau, American Community Survey [9]: This data source provides the percentage of a state's population that have a high school degree or higher, or a bachelor's degree or higher.

U.S. Census Bureau, Population and Housing Unit Estimates [10]: State population numbers are available for each year starting 2000. For years prior to that included in the study (1997-1999), we used raw data from the Intercensal State and County Characteristics Population Estimates and aggregated them up to the state levels.

Federal Election Commission, U.S. Presidential Election Results [11, 12]: We obtained percentage of votes received by Democratic and Republican candidates per state in each presidential election between 2000 and 2016 from this source. For the 1996 presidential election, we processed the raw data available in the ASCII files to obtain the state level vote breakdowns. To estimate the political leaning in the intervening years between presidential elections, we used linear interpolation of the data from two consecutive presidential elections.

Bureau of Economic Analysis (BEA), U.S. Department of Commerce [13]: This data source provided the GDP per state for each year during the study period.

Lawrence Berkeley National Laboratory, Price of PV systems in the US [14]: We obtained installed price of grid-connected, residential and non-residential solar photovoltaic (PV) systems in the US from this report for each year in the study. We assumed the prices to be the same across US due to unavailability of reliable and comprehensive solar cost time series data at state level for the study period.

4. Random Forests

Random forests [15] are an ensemble machine learning technique that constructs many decision trees from the training data and computes the final result by taking a mean of the results predicted by each individual tree. This is an effective way to reduce the problem of overfitting by averaging away any biases in the individual trees. The algorithm used for this research is implemented in Python and is freely available as part of the SciPy module.

4.1. Predicting next year's solar growth

We applied a random forest regressor [15] on our dataset by randomly selecting 75% of the (year, state) data for training and the rest 25% for testing. Using all of the 22 features, the model's efficacy (measured by the R^2 value) is around 90% (0.901). The importance score of the different independent variables is given in Figure 1. The feature serial numbers on the x-axis are taken from Table 1. From these results, this year's solar growth seems to be the strongest predictor of next year's growth.

4.2. Modeling current year solar growth using independent variables

We also constructed a random forest model to predict this year's solar growth based on other features to identify the strongest factors that explain the growth in solar energy within a state. Using 19 features, the model's efficacy is around 86% (0.862). The relative importance of features is shown in Figure 1.

5. Regression techniques

We utilized three different regression techniques to predict values for 1-year-forward solar growth: linear, polynomial, and ridge. Linear regression fits the data points in the model to a line and returns a linear equation that can be used to predict unknown values. Polynomial regression does this for degrees higher than one. Ridge regression is slightly more involved: first, L2 regularization (penalty equivalent to the square of the magnitude of coefficients) is performed. Next, a λ (tuning parameter) is specified. The value of λ determines the amount of emphasis given to minimizing the sum of the square of coefficients. For models with $\lambda = 0$, ridge regression outputs the same coefficients as simple linear regression. And for models with $\lambda = \infty$, the coefficients will all be zero due to the infinite weightage on the square of coefficients [16]. We used Python implementations from the scikit-learn library.

Table 1: Factors used in developing our models for explaining current year's solar growth and predicting next year's solar growth

Serial No.	Feature Name	Serial No.	Feature Name
1	Solar potential of the state	12	Cost of solar installation
2	Democratic vote / Republican vote	13	Growth in cost of solar installation
3	Total population	14	This year's growth in Democratic vote / Republican vote
4	Median household income	15	This year's population growth
5	GDP	16	Growth in median household income
6	% of population with a bachelor's degree or higher	17	GDP growth %
7	% of population with a high school degree or higher	18	Total energy produced in the state
8	% of population that is white	19	This year's growth in total energy produced
9	% of population that is black	20	Total solar energy consumed in the state
10	% of population that is Hispanic	21	% of energy coming from solar
11	% of population that is Asian	22	This year's solar growth %

IOP Conf. Series: Earth and Environmental Science **127** (2018) 012002 doi:10.1088/1755-1315/127/1/012002

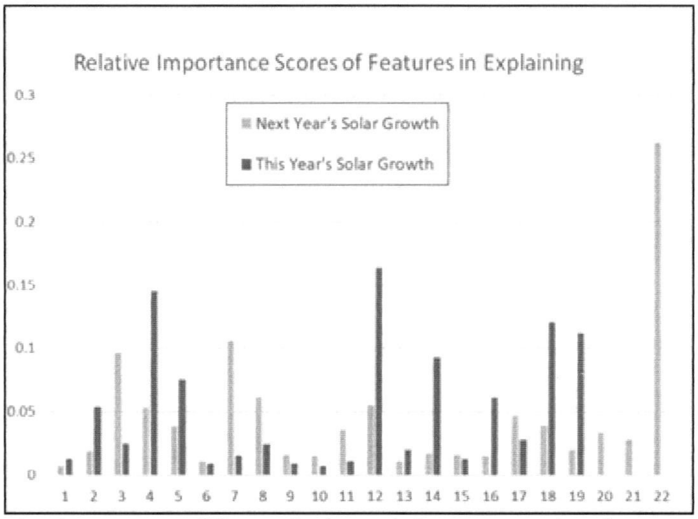

Figure 1: Relative importance of factors in determining next year's solar growth vs. Relative importance of factors in determining this year's solar growth

Using 1-year-forward solar growth % as the dependent variable, we first found the correlation of each independent variable. As anticipated, the most highly-correlated independent variable was this year's solar growth. Using it as the sole independent variable resulted in the model expressed in equation (1) and Figure 2(a) with an R^2 value of 0.1024.

$$y = 0.31959935\, x + 12.18611002 \tag{1}$$

We then used multivariate linear regression, using from 2 to 22 independent variables to predict 1-year-forward solar growth. To do this, we iteratively created every possible combination of parameters in the linear model. Each additional variable increased the R^2 by diminishing margins, shown in Figure 2(b). The R^2 plateaus off at around 10 variables. The highest correlation achieved with linear regression, using all 22 independent variables, was 0.1762. Both polynomial and ridge regression techniques produced poor results with very high error rates. Those results are not included here.

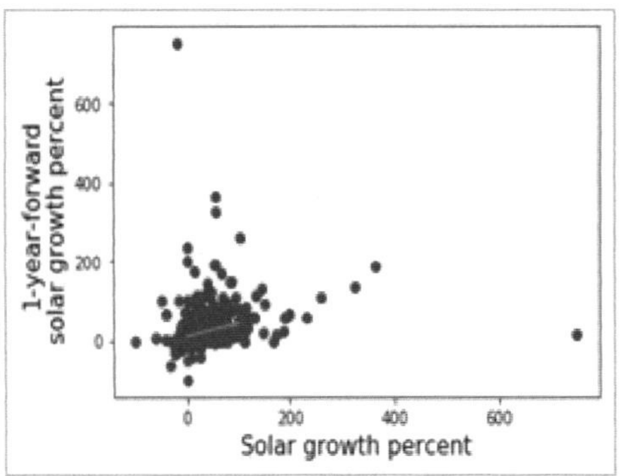

(a)

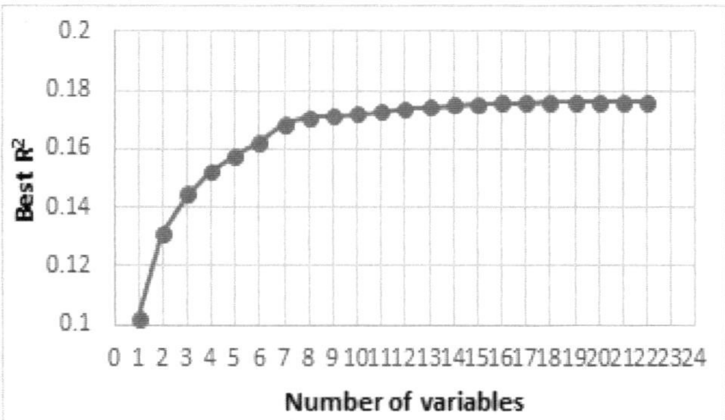

(b)

Figure 2: (a) This year's solar growth vs. Next year's solar growth (b) Increase in R2 with increasing number of explanatory variables

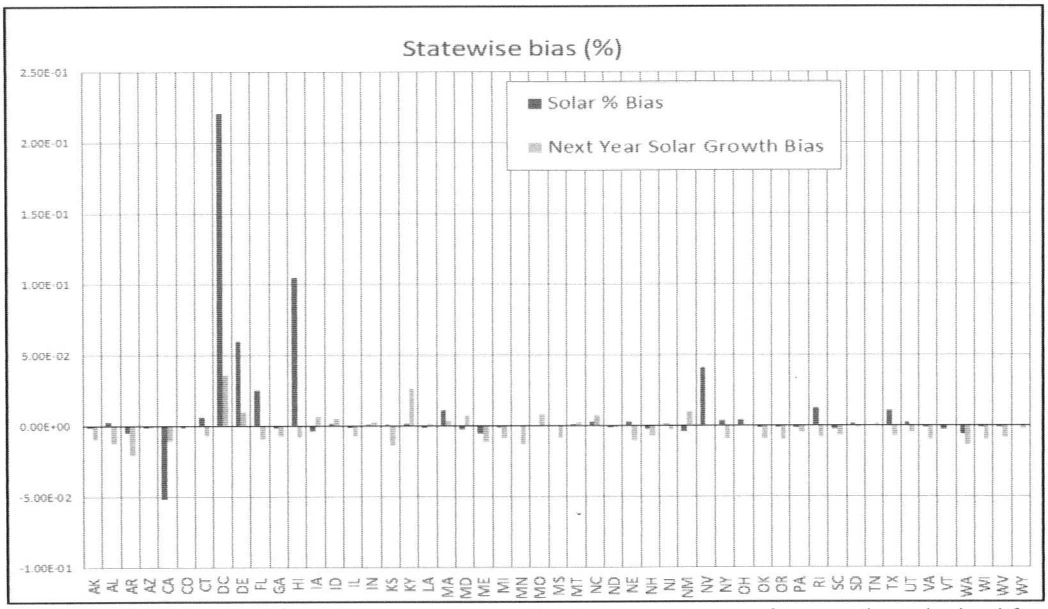

Figure 3. State-wise biases for current year solar % and next year solar growth as obtained from residual analysis of the SVM regressor

6. Support Vector Machines

Support vector machine (SVM) is a machine learning tool first introduced by Vapnik et al [17]. In SVM regression, the input **x** is first mapped onto an m-dimensional feature space using a nonlinear mapping, and then a linear model is constructed in this feature space. The SVM regressor used in this work is the one distributed with MATLAB. The linear model in the feature space $f(x,w)$ is given by equation (2) where $g_j(x)$, $j=1,\ldots,m$ denotes a set of nonlinear transformations, and b is the 'bias' term [18].

$$f(x,w) = \sum_{j=1}^{m} w_j g_j(x) + b \tag{2}$$

We randomly selected about 75% of the years and used the data from these years for training an SVM and the remaining years for testing. We found the models generated by our program had a low R^2 value, calculated at less than 0.30 for predicting next year's solar growth from the current year's data, and less than 0.60 for predicting the current year's solar production from the other factors for the current year. To understand these results better, we ran two different sets of experiments as explained below.

6.1. Residual analysis by state

We fitted an SVM regressor to all of the data, and calculated the model residuals for each state for each training year. Then we averaged these residuals for each state over all training years to find the mean residual for each state. These results are shown in Figure 3. We see particularly large variation in the residuals for the current year's solar % by state, which we call state-wise "bias". A negative bias for a state indicates that the state is performing poorer than expected on the solar energy production front, and so, that state has a larger potential for growth. A positive bias, on the other hand, means a state is already performing better than expected and there is a smaller potential for higher growth.

On inspecting the biases, we can find some intuitive understanding for some of the states with the highest bias absolute values. For instance, the highest positive bias is for the District of Columbia which may be due to its small size (smallest among all 'states'), or high degree of urbanization (100%), or some other factor not captured by our data. The second highest positive bias is for Hawaii which may be because it is an island without easy access to sources of conventional energy.

6.2. Residual analysis by year

We perform another set of experiments where we fit an SVM regressor to the data and find the mean residuals for each year averaged over all states. We found the year-wise solar % bias to follow a nearly monotonically increasing trend over the years. This is likely due to some factors not considered in our models, such as changing government policy, or increasing focus on environment, energy security, etc. We plan to analyze these biases further in the future to better understand their cause.

7. Conclusion and future work

In this paper we have tried to solve the hitherto unaddressed problem of predicting the solar energy production and next year's solar growth at a US state level. Our contribution with this research is threefold. First, we consolidate data from over a dozen different datasets and identify factors that have the potential to influence solar energy. Second, we perform a comparative evaluation of different machine learning and regression models for solar growth and identify a random forest regression model that can predict/explain the dependent variables with 90% efficacy. Other models, including linear, polynomial, and ridge regression, were also tested, yielding mixed results. Finally, we analyze the residuals obtained by fitting an SVM regressor on the data and identify state-wise and year-wise biases. The state-wise biases can help us identify states with high or low potentials for growth. In future, we wish to continue analyzing these biases to better understand their cause and improve our models further.

8. References

[1] U.S. Energy Information Administration 2017 Annual energy outlook 2017 with projections to 2050.
[2] Sherwood L, Interstate Renewable Energy Council 2013 U.S. solar market Trends.
[3] Office of Energy Efficiency and Renewable Energy, United States Department of Energy 2012 *Renewable Energy Data Book.*
[4] U.S. Energy Information Administration 2016 State Energy Data System *2016 updates by energy source.*
[5] US Census Bureau 2015 Median household income by state: historical income tables
[6] The Henry J. Kaiser Family Foundation 2015 State health facts- population distribution by race/ethnicity
[7] Renewable Resource Data Center, National Renewable Energy Laboratory 2006 Sun index
[8] Nebraska Government 2006 Comparison of solar power potential by state http://www.neo.ne.gov/statshtml/201.htm
[9] U.S. Census Bureau 2010 *American Community Survey 1-Year Estimates: Education Attainment for States, Percent with High School Diploma and with Bachelor's Degree*
[10] U.S. Census Bureau 2015 *Population and Housing Unit Estimates*
[11] Federal Election Commission 2012 *Election Results for the U.S. President, the U.S. Senate, and the U.S. House of Representatives*
[12] Wasserman D 2016 National popular vote tracker, *Cook Political Report.*
[13] Bureau of Economic Analysis, U.S. Department of Commerce 2015 *Regional Economic Accounts – Annual Gross Domestic Product by State*
[14] Barbose G L and Darghouth N R 2016 Tracking the sun IX: the installed price of residential and non-residential photovoltaic systems in the United States, *Lawrence Berkeley National Laboratory Report*
[15] Ho T K 1995 Random decision forests, *Proc. 3rd Int. Conf. on Document Analysis and Recognition (Montreal)*
[16] Tibshirani R 2013 *Modern regression 1: ridge regression* http://www.stat.cmu.edu/~ryantibs/datamining/lectures/16-modr1.pdf
[17] Vapnik V 1995 *The Nature of Statistical Learning Theory* (New York: Springer).
[18] Smola A and Schölkopf B 1998 A Tutorial on Support Vector Regression, *NeuroCOLT Technical Report NC-TR-98-030* (Royal Holloway College, University of London, UK).

A Method of Dynamic Extended Reactive Power Optimization in Distribution Network Containing Photovoltaic-Storage System

Wu Wang[1], Wei Huang[1], Yongjun Zhang[1]

[1]School of Electrical Power, South China University of Technology, Guangzhou, 510640,China

277126159@qq.com

Abstract. The grid-integration of Photovoltaic-Storage System brings some undefined factors to the network. In order to make full use of the adjusting ability of Photovoltaic-Storage System (PSS), this paper puts forward a reactive power optimization model, which are used to construct the objective function based on power loss and the device adjusting cost, including energy storage adjusting cost. By using Cataclysmic Genetic Algorithm to solve this optimization problem, and comparing with other optimization method, the result proved that: the method of dynamic extended reactive power optimization this article puts forward, can enhance the effect of reactive power optimization, including reducing power loss and device adjusting cost, meanwhile, it gives consideration to the safety of voltage.

1. Introduction

Recent years, the short of traditional energy and the gradual opened of power market, makes the Distributed Generator (DG) get a large number of applications. As the penetration of DG become higher in power network, distribution network will be more complex. Therefore, scholars do a lots research in relevant content, and the problem of reactive power optimization become a hot content [1,2,3].

Dynamic reactive power optimization is a complex mixed-integer nonlinear programming problem, and it commonly increase the maximum action frequency of control variable as constraints to avoid control equipment frequent operate [4]. So far, distributed energy storage technology receive a lots of attentions and has made considerable development [5]. The Energy Storage System (ESS) can reduce the cost and increase the flexibility of the system[6]. ESS offers an ideal economic dispatch[7], If ESS)combined with photovoltaic (PV) connect to the grid, the active power of ESS will be able to two-way adjust, effectively relieve the problem of voltage over limit and equipment overload after distributed photovoltaic access, it can act as a buffer in power system [8]. From the control point of view, if ESS be used in distributed network, ESS's active power can be treated as control variable and it will be able to enhance the adjusting ability of network. Therefore, installing Photovoltaic-Energy Storage Hybrid System (PESHS) in the operation of distribution network as additional will become the main trend.

For this reason, towards distributed network containing PESHS, this paper puts forward a kind of dynamic reactive power optimization model. This model add the minimum adjusting cost of ESS as one of objective, use 24-hour as an optimization period, and optimizing in chronological order.

IOP Conf. Series: Earth and Environmental Science **127** (2018) 012003 doi:10.1088/1755-1315/127/1/012003

2. Mathematical Model of Dynamic Reactive Power Optimization

2.1. Problem formulation

The proposed model adopts the method of time decoupling, the dynamic reactive power optimization is transformed into static reactive power optimization.

(1) Minimum active loss of network

Active loss is an important index of power grid operation economical efficiency, adjusting the state of transformer and reactive power equipment, or the active and reactive output of PESHS, the load flow distribution would be improved, so as to effectively reduce the active loss of power grid. Its mathematical expression is as follows:

$$P_{\text{LOSS}} = \sum_{i=1}^{n} V_i \sum_{j\in\Gamma} V_j \left(G_{ij} \cos\delta_{ij} + B_{ij} \sin\delta_{ij} \right) \tag{1}$$

(2) Minimum adjusting cost of transformer and compensating capacitor

$$C_{\text{T-Q}}^t = C_{\text{T}} \sum_{i=1}^{n_{\text{T}}} (T_i^t - T_i^{t-1}) + C_{\text{Q}} \sum_{j=1}^{n_{\text{Q}}} (Q_j^t - Q_j^{t-1}) \tag{2}$$

Where t=time t; i= transformer i; j= compensating capacitor j; $C_{Q\text{-}T}$ is the total adjusting cost of transformer and compensating capacitor ; C_{T} and C_{Q} are the unit adjusting cost of transformer and compensating capacitor; T is transformer tap notch and, Q is group number of operational capacitor; n_{T}, n_{Q} are number of transformer and compensating capacitor.

(3) Minimum adjusting cost of ESS

$$C_{\text{E}}^t = \sum_{i=1}^{n_{\text{E}}} C_{\text{OMB}} (P_{\text{C}i}^t + P_{\text{D}i}^t) + C_{\text{Dep}} U_{\text{D}i}^t (1 - U_{\text{D}i}^{t-1}) \tag{3}$$

Where t=time t; i=ESS i; C_{E}^t is the operating cost of ESS; C_{OMB} is the maintain cost coefficient of ESS; $P_{\text{C}i}^t$, $P_{\text{D}i}^t$ are the charging and discharging power of ESS, $U_{\text{D}i}^t$ is the state of charge; C_{Dep} is the ESS's cost of depreciation, as in (4):

$$C_{\text{Dep}} = \frac{\eta_{\text{Dep}}}{N} (C_{\text{P}} P_{\text{N}} + C_{\text{E}} E_{\text{N}}) \tag{4}$$

Where η_{Dep} is the depreciation coefficient of ESS; P_{N} is the maximum output of ESS; E_{N} is rated capacity of ESS; C_{P} and C_{E} are the corresponding installing cost coefficient, N is cycle index of accumulator. In order to make units unified, using electrical power form to express adjusting cost of ESS, as in (5):

$$C_{\text{E}}^{t\,\prime} = \frac{C_{\text{E}}^t}{\alpha \Delta t} \tag{5}$$

Where α is electricity price; Δt is unit time of optimization.

Synthesizing these optimization goals above, the objective function can be concluded:

$$\min F^t = P_{\text{LOSS}}^t + C_{\text{T-Q}}^t + C_{\text{E}}^{t\,\prime} \tag{6}$$

Control variable include:

Q_{C}=reactive power output of compensating capacitor;

T_{K}=voltage regulator;

λ_{PV}=power factor of photovoltaic system;

P_{E}=active power output of ESS;

2.2. Constraint condition

The objective function is restricted by the following factors.

Limit on unit capacitor capacity:

$$Q_{C\min} < Q_{Ci} < Q_{C\max} \tag{7}$$

Limit on tap position of transformer:

$$T_{K\min} < T_{Ki} < T_{K\max} \tag{8}$$

Limit on PSS active power output:

$$[V_{Di} + P_{Di}] < V_{Gi} < [V_{Di} + P_{Ui}] \tag{9}$$

Voltage magnitude limits at PQ nodes:

$$0.95 \le V_{pq} \le 1.05 \tag{10}$$

Power factor limit of PV generation:

$$\lambda_{\min} \le \lambda_{PVi} \le \lambda_{\max} \tag{11}$$

Where $Q_{C\min}$, $Q_{C\max}$ are the minimum and maximum capacity of unit capacitor i; $T_{K\min}$, $T_{K\max}$ are the minimum and maximum tap position of transformer i; $\lambda_{\min}$, $\lambda_{\max}$ are the minimum and maximum power factor of photovoltaic; V_{Di} is the active power output of photovoltaic i; P_{Ui} and P_{Di} are the minimum and maximum active power output of ESS i, they can be obtained as:

$$\begin{cases} P_{Ui} = \min(\dfrac{E_{\max}}{\eta_{Ci}\Delta t}, \dfrac{E_{Ui} - E_i}{\eta_{Ci}\Delta t}) \\ P_{Di} = \max(E_{\min}\dfrac{\eta_{Di}}{\Delta t}, (E_{Di} - E_i)\dfrac{\eta_{Di}}{\Delta t}) \end{cases} \tag{12}$$

Where Δt is the unit time of optimization; E_{Di}, E_{Ui} are the minimum and maximum capacity of ESS i; $E_{\min}$, $E_{\max}$ are the maximum discharge-charge power network receive from ESS; η Ci, η Di are the efficiency of charge-discharge power; E_i is the remainder capacity of ESS, the calculating method can be described as in (13):

$$E_i^t = E_i^{t-1} + (P_{C,i}^t \eta_{C,i} - \frac{P_{D,i}^t}{\eta_{D,i}})\Delta t - E_1(1 - U_{C,i}^t)(1 - U_{D,i}^t) \tag{13}$$

Where t=time t; P_{Ci}/η Ci, P_{Di}/η Di are the charge-discharge power of ESS; U_{Ci}, U_{Di} are the state of ESS charge; E_1 is the wastage of the energy storage resting.

3. Calculating method

This paper uses Cataclysmic Genetic Algorithm (CGA) to solve reactive power optimization. CGA based on genetic algorithm and adds catastrophe operator, so as to improve the efficiency of optimization, and avoid the optimization calculation into the local optimum. Using MATLAB to write relevant program, and is used to optimize calculation.

In order to satisfy the voltage constraint, this paper use a Voltage Risk Index, it is regard as a penalty term join in the objective function, it is obtained as:

$$V_{LIM} = \sum_{i=1}^{n} \frac{|V_i - Sat(V_i)|}{V_{\max,i} - V_{\min,i}} \tag{14}$$

Where V_i is magnitude of Voltage on node i; $Sat(x)$ is saturation function, it can be obtained as:

$$Sat(x) = \begin{cases} x_{\min}, x < x_{\min} \\ x, x_{\min} \le x \le x_{\max} \\ x_{\max}, x_{\max} < x \end{cases} \tag{15}$$

Meanwhile, in order to make the energy of ESS balance between the beginning and end of optimization period, this paper use Storage Energy Balance Index to be regard as a penalty term appended in objective function, it can be obtained as:

$$E_{\mathrm{cd}i}^{t}=\begin{cases}E_i^t-E_i^1, & (E_i^t-E_i^1)>\varepsilon\\ 0, & (E_i^t-E_i^1)\le\varepsilon\end{cases} \tag{16}$$

Where E_i^t is the remainder capacity of ESS i at the time of t, ε is value of allowable deviation.

After adding penalty terms to objective function, the extended objective function is:

$$\min F^{t\prime\prime}=P_{\mathrm{LOSS}}^{t}+C_{\mathrm{T\text{-}Q}}^{t}+C_{\mathrm{E}}^{t\prime}+\lambda_{\mathrm{V}}V_{\mathrm{LIM}}^{t}+\lambda_{\mathrm{E}}\sum_{i=1}^{n_{\mathrm{E}}}E_{\mathrm{cd},i}^{t} \tag{17}$$

Where λ_{V}, λ_{E} are penalty divisor, and λ_{V} is a positive number, the value of λ_{E} increase gradually, it can be obtained as:

$$\lambda_{\mathrm{E}}=\begin{cases}0, t\le t_1\\ k_1\cdot t^2, t_1<t\le t_2\\ k_2\cdot t, t_2<t\le T\end{cases} \tag{18}$$

In order to ensure that λ_{E} increase gradual, equation (18) is subject to $k_1<k_2/t_2$.

4. Test systems

This paper uses a distributed network as an example, the topological structure as shown in Fig.1, it has 108 nodes. Forecast of load rate as shown in Fig.2, 3 p.m. reach the maximum load 6.246MW. There are three peak load periods: 9 a.m. -11 a.m., 1 p.m. - 5 p.m. and 8 p.m. - 10 p.m.

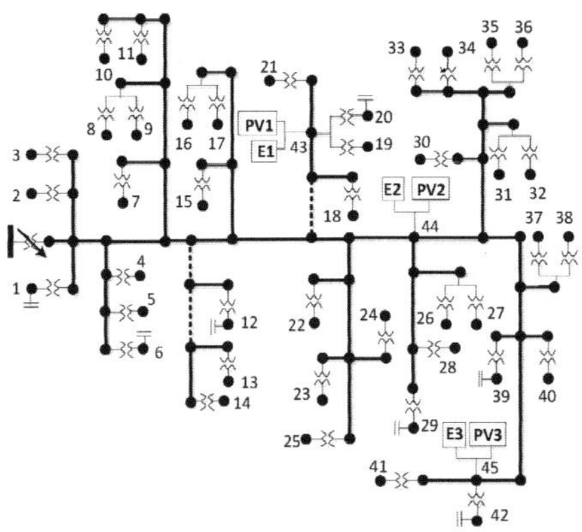

Figure 1. A topological structure of one distributed network.

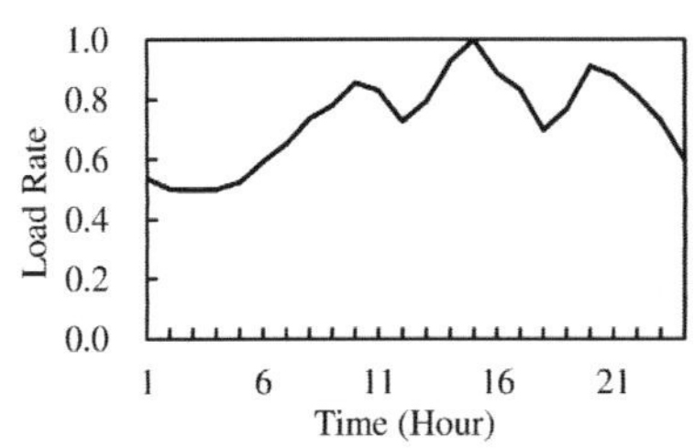

Figure 2. Forecast of load rate.

There are three PESHS access network through nodes 43, 44, 45 respectively, the configuration and index of PESHS shown in Table 1. A voltage regulator in the power system contains 17 tap position （UN±8×1.25%, capacitors installed in node 4, 10, 16, 22 ,25, 28, 33, capacity of unit group are 20kvar, 10kvar, 10kar, 10kvar, 10kvar, 15kvar, 75kvar, all capacitors have 2 groups to operate.

Table 1. Configuration and Index of PESHS

Index	Parameter	Index	Parameter
E_N	$3\text{MW}\cdot\text{h}$	C_p	667CNY / kW
P_N	800kW	C_e	$1200\text{CNY} / (\text{kW}\cdot\text{h})$
η_c	85%	C_{omb}	0.01CNY / kW
η_d	85%	η_{dep}	0.1%
E_l	$2(\text{kW}\cdot\text{h})/\text{h}$	α	$0.8\text{CNY}/\text{kW}\cdot\text{h}$
$E_{\text{D}i}$	$0.6\text{MW}\cdot\text{h}$	C_T	3kW
$E_{\text{U}i}$	$3\text{MW}\cdot\text{h}$	C_Q	2kW

Installed capacity of PV1, PV2 and PV3 are 350kW, 300kW and 250kW, range of power factor are all from lag 0.85 to lead 0.85. Fig.3 displays the 24 hours forecast for the PV power output.

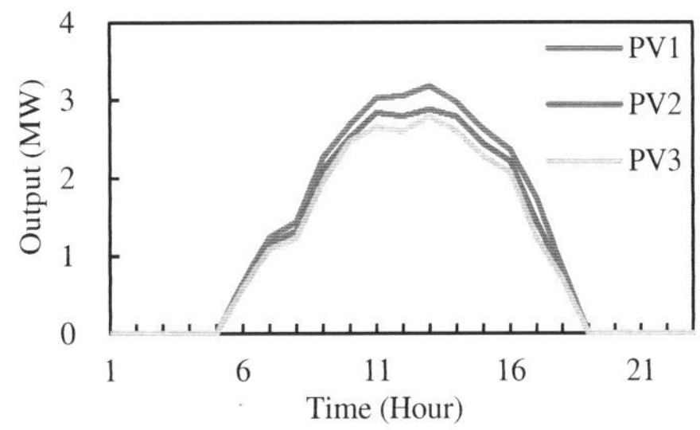

Figure 3. Forecast for the PV power output.

Different optimization method has been set so as to find different optimization effect. In order to compare the effects of the reactive power output of photovoltaic inverter and energy storage on distribution network operation, the control variable of each method as shown: Method 1: Q_C, T_k; Method 2: Q_C, T_k, λ_{PV}; Method 3: Q_C, T_k, λ_{PV}, P_{ES}. By using CGA, we can calculate the optimal solution and get the results. Objective value of different method are shown as Fig.4, and power loss is shown as Fig.5.

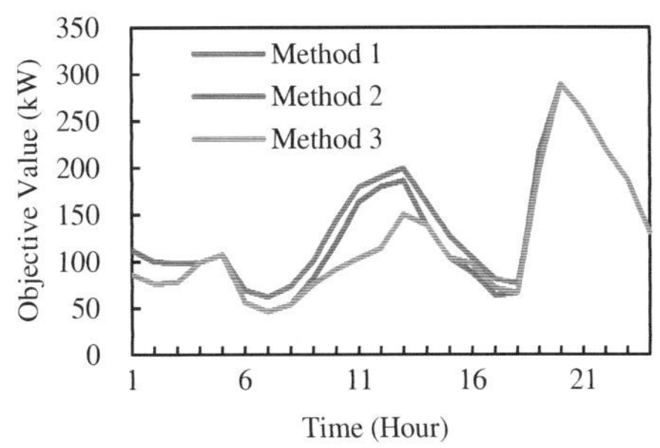

Figure 4. Objective value.

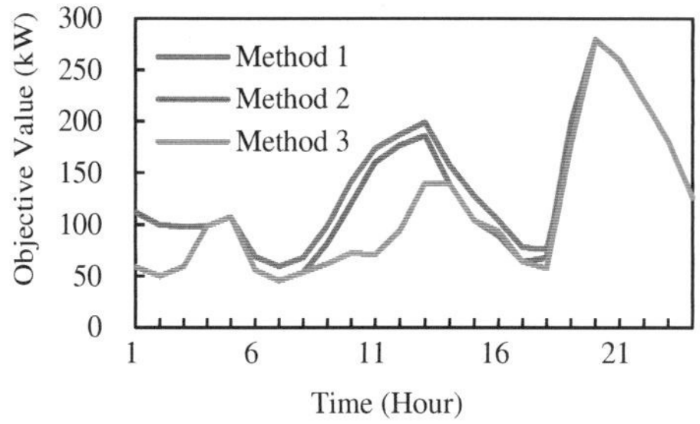

Figure 5. Power loss.

According to Fig.4 and Fig.5, comparing method 1 with 2, we can see that after the photovoltaic inverter provides reactive power, power loss is obviously reduced. Relative to methods 1 and 2, method 3 is the best optimization method. Because of the enough original energy of ESS, in the first several time ESS provide active power to reduce power loss. At noon time, ESS absorb extra photovoltaic power and reduce power loss. In the remaining time, due to energy storage charge and discharge balance constraints, in order to ensure the normal operation of the next optimization cycle, energy storage device in keeping the capacity close to the initial capacity and then tend to stop run, so the optimization result is the same as the other methods.

Sum up the 24 hours calculation results, total objective value and total power loss of different method as shown in the Table 2.

Table 2. Comparison in Total Objective Value and Total Power Loss

Method	Total Objective Value (MW)	Improve Ratio	Total Power Loss (MW)	Improve Ratio
1	3.3614	—	3.3184	—
2	3.1634	5.89%	3.1264	5.79%
3	2.8963	13.84%	2.6693	23.4%

Compare method 2 with 1, the former can reduce more objective value and power loss. According to the optimization effect of method 3, the combined control of reactive power output of photovoltaic inverter and storage energy can achieve better optimization results. Therefore, in the reactive power optimization which consider PESHS, add active power output of ESS and power factor of PV unit as the expansion control variable, better results can be achieved in the grid economy.

5. Conclusion

This paper puts forward an extended reactive power optimization model, which based on PESHS. The model via time decoupling to turn dynamic reactive power optimization to static reactive power optimization. In the optimization, the power loss and the adjustment cost of the equipment are taken into consideration. By setting different optimization methods, we can draw the following conclusions:

1. The reactive power provided by the PV inverter can effectively reduce power loss and total objective value without any other cost.

2. ESS can improve the ability to regulate the distribution network. In a distribution network with photovoltaic systems, ESS can provide enough active power to make up for the shortage of photovoltaic power, and can absorb excess PV active power. Although this may increase the operating cost of the equipment, it is superior in overall economic performance.

3. PESHS access to the distribution network will increases the controllable variables, improves the reactive power control capability of the distribution network, and further excavate the potential of energy saving.

The research done in this paper can be further improved, the influence of PV access uncertainty on active distribution network can be considered, and the effect of ESS on regulating voltage can be further studied. These will be improved in the future research.

6. References

[1] T. Ding, S. Liu, W. Yuan, et al, "A Two-Stage Robust Reactive Power Optimization Considering Uncertain Wind Power Integration in Active Distribution Networks," *IEEE Transactions on Sustainable Energy*, vol. 7, Nov. 2015, pp. 301-311, doi:10.1109/TSTE.2015.2494587.

[2] W. Sheng, K. Liu, Y. Liu, et al, "Reactive power coordinated optimization method with renewable distributed generation based on improved harmony search." *IET Generation, Transmission & Distribution*, vol. 10, Oct. 2016, pp. 3152-3162, doi:10.1049/iet-gtd.2015.1051.

[3] W. Zheng, W. Wu, B. Zhang, et al, "A Fully Distributed Reactive Power Optimization and Control Method for Active Distribution Networks," *IEEE Transactions on Smart Grid*, vol. 7, Feb. 2015, pp. 1021-1033, doi:10.1109/TSG.2015.2396493.

[4] G. Geng, J Liang, R. G. Harley, and R. Qu, "Load profile partitioning and dynamic reactive power optimization," *2010 International Conference on Power System Technology*, Oct. 2010, pp. 1-8, doi:10.1109/POWERCON.2010.5666514.

[5] Y. Xu, W. Zhang, G Hug, et al, "Cooperative Control of Distributed Energy Storage Systems in a Microgrid," *IEEE Transactions on Smart Grid*, vol.6,Sep. 2014, pp. 238-248, doi: 10.1109/TSG.2014.2354033.

[6] Y. Zhou, L. Zhang, J. Mutale, et al, "System level assessment of PV and energy storage: Application to the Great Britain power system," *2017 IEEE Manchester PowerTech*, June 2017, pp:1-6, doi: 10.1109/PTC.2017.7981270.

[7] X. Shen, M. Shahidehpour, Y. Han, et al, "Expansion Planning of Active Distribution Networks With Centralized and Distributed Energy Storage Systems," *IEEE Transactions on Sustainable Energy*, vol. 8, June. 2016, pp. 126-134, doi:10.1109/TSTE.2016.2586027.

[8] J. H. Teng, S. W. Luan, D. J. Lee, and Y. Q. Huang, "Optimal Charging/Discharging Scheduling of Battery Storage Systems for Distribution Systems Interconnected With Sizeable PV Generation Systems," *IEEE Trans. Power Syst.*, vol. 6, no. 2, pp. 726-736, March 2015.

Acknowledgment

The authors gratefully acknowledge the funding support by the Key Program of the National Natural Science Foundation of China (51377060).

IOP Conf. Series: Earth and Environmental Science **127** (2018) 012004 doi:10.1088/1755-1315/127/1/012004

Mismatch locating strategy of photovoltaic array

R X Fan[1], J R Miao[2], C Zhao[3] and D Xie[*,2]

[1] Institute of Electric Power Research, State Grid Jiangxi Electric Power Company, Nanchang, Jiangxi 330096, China;
[2] School of Electronic Information and Electrical Engineering, Shanghai Jiao Tong University, Minhang District, Shanghai 200240, China
[3] School of Electrical Engineering, Shanghai Dianji University, Minhang District, Shanghai 200240, China

[*]E-mail: xieda@sjtu.edu.cn

Abstract. Single-diode model of Photovoltaic cell is adopted in this paper and is modified by adding a mismatch reflect parameter. The deduced mathematical model of serial PV is provided in this paper. The unknown parameters in the model is determined by the environment conditions including the irradiation and the temperature. With the measured data of the current and voltage values, equation set about the unknown quantities can be listed out. Then iteration method is introduced to solve them. The mismatch reflect parameter is vital to locate the mismatched series in the PV array.

1. Introduction

With the development of photovoltaic generation system, more and more researches are concentrated on it. Before, a large proportion of attention is paid on the improvements of the efficiency. Many novel MPPT strategies are proposed to increase it. Researches on the mismatch problem are quite rare in compare [1]. Mismatch occurs when the electrical property, such as voltage, current, power, between components in the cell array is quite different. Thus, the output power of the entire PV array will be less than the sum of the individual module. There are many causes leading to the phenomenon including internal causes such as mixed connection of PV components with different output power, manufacturing differences and the aging problems and external causes such as the different orientation of sunlight, partial cover by the cloud shadow [2].

The most common way to alleviate mismatch when the mismatch occurs, the output power of the system will be affected. Meanwhile, the normal components will also have efficiency loss because of voltage or current constraints. If a single cell or only a minority of cells are mismatched because of shadow or other cases, the mismatched cells will get locally overheated, which will cause “hot spot effect”. This phenomenon will accelerate the aging and damage of PV cells [3].

The mismatch problem is mainly alleviated by adding the bypass diode to compensate for current mismatch or adding the blocking diode to compensate for voltage mismatch. The current solution for “hot spot effect” is to connect the diode in parallel, which is called the bypass diode. When part of the cells in the same bypass diode are mismatched, current of the cells which are parallel to the same bypass diode will directly flow through the bypass diode. Thus, the voltage is limited and “hot spot effect” is eased. However, the bypass diode can only alleviate the hot spot phenomenon instead of completely eliminating its damage [4].

In [5], DC resistance of the serial module instead of bypass diode is used to detect hot spot. Then "hot spots" are cut off by biostable relay. The cut-off part can be re-connected into the circuit after the temporary mismatch. However, this method will still cause power loss in the same series and the equipment cost is high. In [6], the internal conditions and external conditions of hot spot failure considering the bypass diode in parallel are obtained. The quantity of maximum cells that can be connected with the same bypass diode in series is induced via the simplified U-I model. While this method can be only applied to a small system containing one or two bypass diodes. Also, this method is based on the current and voltage values of the single PV cell and is difficult to achieve in reality.

A mismatch locating strategy is proposed in this paper. Part Ⅰ is the introduction. The single-diode model of Photovoltaic cell and its modification is introduced in part Ⅱ; the deduced mathematical model of serial PV is also discussed in this part. Part Ⅲ introduces the approach to solving the parameters in detail and part Ⅳ is the application and result. The last part is discussion. The major contributions of this paper are:

(1) It proposes an approach to locate the mismatch series of PV array. When mismatch is caused by internal causes, actions can be taken to directly find out the fault series and solve it.

(2) The location algorithm is based on mechanism model. Thus, it is applicable to the mismatch circumstances whatever induced by internal or external causes.

2. Mathematical model

2.1. Mathematical model of PV cell

The single-diode model of PV cell is shown in Fig.1.

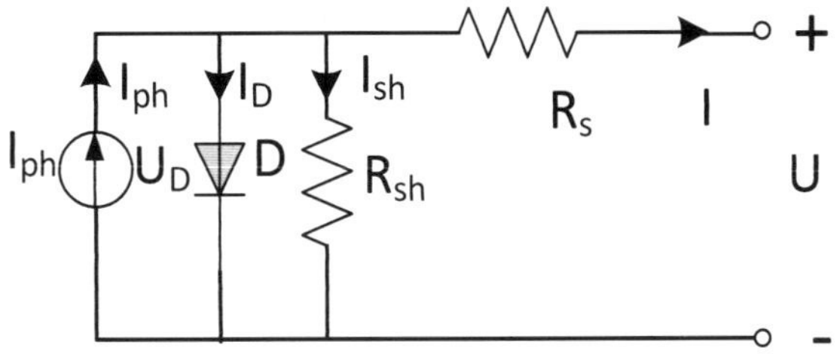

Figure 1. The equivalent circuit of photovoltaic cell

The curve equation of the output voltage and current is shown in (1):

$$I = I_{ph} - I_s\left(e^{\frac{q(U+IR_s)}{AkT}} - 1\right) - \frac{U+IR_s}{R_{sh}} \tag{1}$$

Where q is the electron charge constant, 1.602 × 10-19 C; k is the Boltzmann constant, 1.381 × 10-23 J / K; A is the diode characteristic fitting coefficient, which is variable between 1 and 2. This relationship can be seen as an equation with five parameters to be determined, namely Iph, Rsh, a, Is and Rs. Iph is the photo-generated current; q/AkT is denoted by a here; Is is the saturation current of equivalent diode; Rsh is the equivalent parallel resistance; Rs is the equivalent series resistance.Some text.

2.2. Modified formulas of U-I equation

Modified formulas of Iph, Rsh, Is and Rs based on irradiation and temperature values are shown in (2):

$$\begin{cases} I_{ph} = \left(\dfrac{S}{\mathrm{S_{ref}}}\right)\left[\mathrm{I_{phref}} + \mathrm{C_T}\left(T - \mathrm{T_{ref}}\right)\right] \\ I_s = \mathrm{I_{sref}}\left(\dfrac{T}{\mathrm{T_{ref}}}\right)^3 \mathrm{e}^{\left[\frac{\mathrm{qE_g}}{kA}\left(\frac{1}{\mathrm{T_{ref}}} - \frac{1}{T}\right)\right]} \\ R_s = \mathrm{R_{sref}}\left(\dfrac{T}{\mathrm{T_{ref}}}\right)\left[1 - \beta\ln\left(\dfrac{S}{\mathrm{S_{ref}}}\right)\right] \\ R_{sh} = \mathrm{R_{shref}}\left(\dfrac{S}{\mathrm{S_{ref}}}\right) \end{cases} \tag{2}$$

Where Iphref is the photo-generated current under standard condition(SC). S is the real-time irradiation (W / m2). Sref is the SC irradiation, 1000W/m2. CT is the temperature coefficient which can be provided by the manufacturer. Eg is the forbidden band width (eV), depending on the material of the photovoltaic cell. T is the real time temperature of the diode. Tref is the SC temperature, 25℃. β is a constant taken as 0.217.

2.3. U-I relationship of serial modules

Suppose that a PV module consists of 3 units and each unit consists of 20 cells. Each unit is connected with a bypass diode in parallel. As is shown in Figure 2.

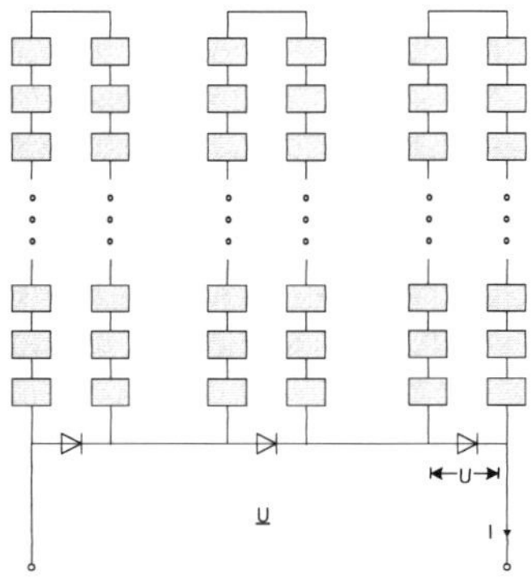

Figure 2. Schematic diagram of PV module

Modules in series can be seen as a module with several units. As is shown in Figure 3.

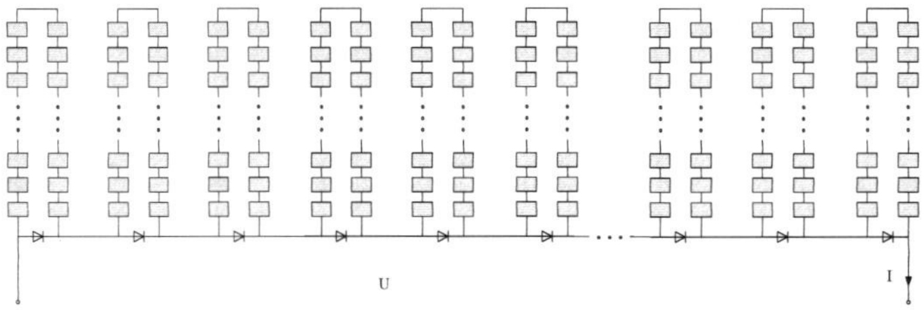

Figure 3. Schematic diagram of PV modules in series

When part of units in a module series are masked, the current will pass directly through the bypass parallel diode. The voltage of each group will be limited to the same as the voltage of the diode, which is 0.3V for the silicon diode denoted by Ud. Considering a PV array with n modules in each series, and totally m series in parallel. As is shown in Figure 4.

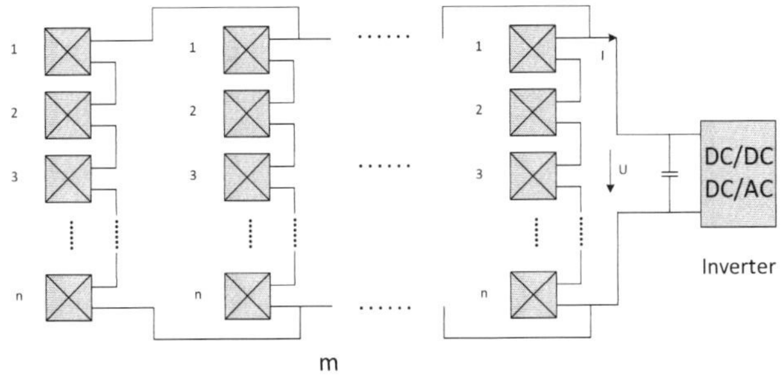

Figure 4. Sketch map of PV array

Assume that there are n1 cells are un-mismatched and n2 mismatched, the sum of n1 and n2 is 3n. Cell voltage considering mismatch is shown in Equation (3):

$$U_D = \frac{U}{n_1} - \frac{n_2 \mathrm{U_d}}{20n_1} + IR_s \tag{3}$$

So, the equation of the output voltage and current of the module series considering mismatch is shown in (4). The relationship of output voltage and current is non-linear and contains exponential function.

$$I = I_{ph} - I_s\left(e^{a\mathrm{U_D}} - 1\right) - \frac{\mathrm{U_D}}{R_{sh}} \tag{4}$$

3. Solution of the parameters

3.1. The solution of parameters

For a series of PV module, there are 6 quantities to be solved, including *Iph*, *Rsh*, *a*, *Is* , *Rs* and the mismatch reflecting parameter n2. Voltage value and current value can be achieved by real time measurement system. Six sets of voltage and current values are needed to solve all the parameters, denoted by:

(U1, I1), (U2, I2), (U3, I3), (U4, I4), (U5, I5), (U6, I6)

After obtaining the data, the equation set can be achieved as (5):

$$\begin{cases} I_i = I_{ph} - I_s\left(e^{a\mathrm{U}_{\mathrm{D}i}} - 1\right) - \dfrac{\mathrm{U}_{\mathrm{D}i}}{R_{sh}} \\ U_{Di} = \dfrac{U_i}{n_1} - \dfrac{n_2 \mathrm{U_d}}{20n_1} + I_i R_s \end{cases} \quad i = 1,2,3,4,5,6 \tag{5}$$

It can be seen as a equation set with six varieties, namely *Iph*, *Rsh*, *a*, *Is*, *Rs* and n2. Different environment circumstances correspond to different parameter values. In a short duration, the five parameters can be seen as constants. Since the parameters have the upper and lower limits, it is easy to choose the initial value. Thus, it can be solved via Newton iteration method. The zero function corresponding to the six equations is Fi, as is shown in (6).

$$\begin{cases} F_i = I_{ph} - I_s\left(e^{a\mathrm{U}_{\mathrm{D}i}} - 1\right) - \dfrac{\mathrm{U}_{\mathrm{D}i}}{R_{sh}} - I_i \\ U_{Di} = \dfrac{U_i}{n_1} - \dfrac{n_2 \mathrm{U_d}}{20n_1} + I_i R_s \end{cases} \quad i = 1,2,3,4,5,6 \tag{6}$$

The Jacobi matrix is denoted by J, and is as follows:

$$J = \begin{bmatrix} \frac{\partial F_1}{\partial I_{ph}} & \frac{\partial F_1}{\partial I_s} & \frac{\partial F_1}{\partial a} & \frac{\partial F_1}{\partial R_s} & \frac{\partial F_1}{\partial R_{sh}} & \frac{\partial F_1}{\partial n_2} \\ \frac{\partial F_2}{\partial I_{ph}} & \frac{\partial F_2}{\partial I_s} & \frac{\partial F_2}{\partial a} & \frac{\partial F_2}{\partial R_s} & \frac{\partial F_2}{\partial R_{sh}} & \frac{\partial F_2}{\partial n_2} \\ \frac{\partial F_3}{\partial I_{ph}} & \frac{\partial F_3}{\partial I_s} & \frac{\partial F_3}{\partial a} & \frac{\partial F_3}{\partial R_s} & \frac{\partial F_3}{\partial R_{sh}} & \frac{\partial F_3}{\partial n_2} \\ \frac{\partial F_4}{\partial I_{ph}} & \frac{\partial F_4}{\partial I_s} & \frac{\partial F_4}{\partial a} & \frac{\partial F_4}{\partial R_s} & \frac{\partial F_4}{\partial R_{sh}} & \frac{\partial F_4}{\partial n_2} \\ \frac{\partial F_5}{\partial I_{ph}} & \frac{\partial F_5}{\partial I_s} & \frac{\partial F_5}{\partial a} & \frac{\partial F_5}{\partial R_s} & \frac{\partial F_5}{\partial R_{sh}} & \frac{\partial F_5}{\partial n_2} \\ \frac{\partial F_6}{\partial I_{ph}} & \frac{\partial F_6}{\partial I_s} & \frac{\partial F_6}{\partial a} & \frac{\partial F_6}{\partial R_s} & \frac{\partial F_6}{\partial R_{sh}} & \frac{\partial F_6}{\partial n_2} \end{bmatrix}$$

The elements in the matrix is unfolded as (7-1) to (7-6); i is integers from 1 to 6. Udi is as the same as (6).

$$\frac{\partial F_i}{\partial I_{ph}} = 1 \tag{7-1}$$

$$\frac{\partial F_i}{\partial I_s} = 1 - e^{aU_{Di}} \tag{7-2}$$

$$\frac{\partial F_i}{\partial a} = -I_s e^{aU_{Di}} U_{Di} \tag{7-3}$$

$$\frac{\partial F_i}{\partial R_s} = -I_s e^{aU_{Di}} a I_i - \frac{I_i}{R_{sh}} \tag{7-4}$$

$$\frac{\partial F_i}{\partial R_{sh}} = U_{Di} \frac{1}{R_{sh}^{\ 2}} \tag{7-5}$$

$$\frac{\partial F_i}{\partial n_2} = -I_s e^{aU_{Di}} \frac{20U_i - n_2 U_d}{20(3n - n_2)^2} - \frac{20U_i - n_2 U_d}{(3n - n_2)^2 R_{sh}} \tag{7-6}$$

Denote that:

$$x = (I_{ph}, R_{sh}, a, I_s, R_s, n_2)'$$

$$F_Z = (F_1, F_2, F_3, F_4, F_5, F_6)'$$

Then the iteration equation can be written as (8):

$$x(n+1) = x(n) - A_n^{-1} F_{Zn} \tag{8}$$

The mismatch parameter n2 reflects the number of mismatched units. As is displayed in Figure 4, the PV array contains m module series in parallel. When we solve the mismatch reflecting parameter of every certain series, we can know the overall mismatch state of the whole PV array. Moreover, we can take actions to find out the mismatched series and solve the mismatch problem.

3.2. Simplification of calculation

From modification formulas we know that *Is*, *a* and *Rs* are impregnability when irradiation changes. As the influence of temperature change in a short duration on parameters can be neglected, and the irradiation remains almost unchanged in a short duration, *Is*, *a* and *Rs* can be regarded as the fixed value in a short duration; and *Iph*, *Rsh* change as the irradiation changes. At a certain time, the irradiation and temperature are determined, so the U-I curve is also determined. Thus, the five parameters are also deterministic. In order to simplify the calculation, *Is*, *A* and *Rs* can be replaced by the values at the nearest integral time; only the values of *Iph* and *Rsh* are required to be calculated. At integral time, all the five parameters are required to be calculated. Thus, we need only 3 sets of values of voltage and current, denoted by:

(U1, I1), (U2, I2), (U3, I3)

The equation is as same as (6) while i is 1,2 and 3. Using Newton iteration method and Jacobi matrix is as follows:

$$J=\begin{bmatrix} \frac{\partial F_1}{\partial I_{ph}} & \frac{\partial F_1}{\partial R_{sh}} & \frac{\partial F_1}{\partial n_2} \\ \frac{\partial F_2}{\partial I_{ph}} & \frac{\partial F_2}{\partial R_{sh}} & \frac{\partial F_2}{\partial n_2} \\ \frac{\partial F_3}{\partial I_{ph}} & \frac{\partial F_3}{\partial R_{sh}} & \frac{\partial F_3}{\partial n_2} \end{bmatrix}$$

The elements in the matrix is unfolded as (9-1) to (9-3); i is integers from 1 to 3. Udi is as the same as (6).

$$\frac{\partial F_i}{\partial I_{ph}}=1 \quad (9\text{-}1)$$

$$\frac{\partial F_i}{\partial R_{sh}}=U_{Di}\frac{1}{R_{sh}^{2}} \quad (9\text{-}2)$$

$$\frac{\partial F_i}{\partial n_2}=-I_s e^{aU_{Di}}\frac{20U_i-n_2U_d}{20(3n-n_2)^2}-\frac{20U_i-n_2U_d}{(3n-n_2)^2R_{sh}} \quad (9\text{-}3)$$

4. Result

We use the measurement system and data acquisition unit and receive the measurement value of voltage and current from the upper computer. Considering the error caused by the measurement equipment and other accidental error, the data gathering system measure the data every few second and transmit the mean values in every 30 second duration to the upper computer. The cloud shade is simulated by artificial barrier.

The data is acquired from a serial PV system. This can be used to confirm the method in this paper. There are three units in a module, 20 modules in series in these case. Four mismatching circumstances are simulated, with zero or one or two or three modules masked respectively. We obtain the voltage and current values as is shown in Tab.1.

Table 1. Database of current and voltage of PV module

n2		Measurements values of voltage and current					
0	U	234.1	393.2	473.3	564.6	594.5	615.2
	I	6.500	6.495	6.462	6.112	5.708	5.196
3	U	254.3	368.5	495.6	586.9	608.9	618.1
	I	6.499	6.485	6.395	5.118	4.172	3.625
6	U	291.2	405.2	517.5	566.1	579.9	592.4
	I	6.497	6.475	5.997	4.715	3.991	3.127
9	U	145.4	249.5	392.4	502.7	536.1	552.5
	I	6.497	6.497	6.468	5.756	4.646	3.660

Parameters can be solved via the method above based on the data in Tab.1. The unit of I is A, and the unit of U is V. The results are displayed in Tab. 2.

Table 2. Parameter value

No.	Iph (A)	Is (μA)	q/AkT	Rs (mΩ)	Rsh (Ω)	n2
1	6.52	3.45	25.6	10.23	8273	0.17
2	6.49	4.33	26.5	9.23	9234	2.93
3	6.50	3.08	26.3	12.82	7263	6.08
4	6.49	2.90	25.9	7.52	10239	9.14

The 5 parameters values under different mismatch circumstances are very close because the temperature and irradiation conditions remains nearly unchanged. The mismatch parameter can basically reflect the mismatch circumstances if ignoring the error.

5. Conclusion

The mismatch parameter n2 can determine whether the serial modules are mismatched. In reality, n is comparatively smaller than m. When the PV array is very large, which is widely used in centralized PV station, to find out the mismatched series is meaningful. If the mismatch is caused by the cloud shade it will disappear when the shade moves; if it is caused by internal causes, the mismatch parameter will always exist, and actions can be taken to find out the mismatched modules and mend them.

6. References

[1] M. Pavlov, A. Migan-Dubois and V. Bourdin, "Experimental and numerical study of the influence of string mismatch on the yield of pv modules augmented by static planar reflectors," *IEEE Journal of Photovoltaics*, vol.5, 2015, pp. 1686-1691, doi: 10.1109/PVSC.2015.7356176.

[2] Y. Hu, W. Cao and J. Ma, "Identifying PV module mismatch faults by a thermography-based temperature distribution analysis," *IEEE Transactions on Device and Materials Reliability*, vol.14, 2014, pp. 951-960, doi: 10.1109/TDMR.2014.2348195.

[3] X. Li, Y. Li and J. E. Seem, "Detection of internal resistance change for photovoltaic arrays using extremum-seeking control MPPT signals," *IEEE Transactions on Control Systems Technology*, vol.24, 2016, pp. 325-333, doi: 10.1109/TCST.2015.2424857.

[4] Y. H. Ji, D. Y. Jung and J. G. Kim, "A real maximum power point tracking method for mismatching compensation in PV array under partially shaded conditions," *IEEE Transactions on power electronics*, vol.26, 2011, pp. 1001-1009, doi: 10.1109/TPEL.2010.2089537.

[5] F. Wan, X. Wu, F. C. Lee, "Analysis of unified output MPPT control in subpanel PV converter system," *IEEE Transactions on power Electronics*, vol.29, 2014, pp. 1275-1284, doi: 10.1109/TPEL.2013.2262102

[6] C. Olalla, C. Deline and D. Clement, "Performance of power-limited differential power processing architectures in mismatched PV systems," *IEEE Transactions on Power Electronics*, vol.30, 2015, pp. 618-631, doi: 10.1109/TPEL.2014.2312980.

Acknowledgment

This paper is supported by The National High Technology Research and Development of China 863 Program (2015AA050103) and State Grid program of Study on Outdoor Testing and Systematic Evaluation of Key Components of PV Power Generation.

IOP Conf. Series: Earth and Environmental Science 127 (2018) 012005 doi:10.1088/1755-1315/127/1/012005

Investigation on thermal environment improvement by waste heat recovery in the underground station in Qingdao metro

Jianwei Liu[1], Jiaquan Liu[2], Fengyin Wang[1], Cuiping Wang[1*]

[1]Energy Engineering Institute, Qingdao University, Shandong Province, China, 266071

[2]Enviormental and Energy Management Department, School of engineering and applied science, The George Washington University, USA

*Corresponding author's email:cpw_qdu@qdu.edu.cn

Abstract. The thermal environment parameters, like the temperature and air velocity, are measured to investigate the heat comfort status of metro staff working area in winter in Qingdao. The temperature is affected obviously by the piston wind from the train and waiting hall in the lower Hall, and the temperature is not satisfied with the least heat comfort temperature of 16 ℃. At the same time, the heat produced by the electrical and control equipments is brought by the cooling air to atmosphere for the equipment safety. Utilizing the water-circulating heat pump, it is feasible to transfer the emission heat to the staff working area to improve the thermal environment. Analyzed the feasibility from the technique and economy when using the heat pump, the water-circulating heat pump could be the best way to realize the waste heat recovery and to help the heat comfort of staff working area in winter in the underground metro station in north China.

1. Introduction

The metro construction industry is developing very quickly in China; and the metro cities are more and more. But the capital cost, energy consumption and energy conservation potential and energy saving methods are causing increasingly attention. The energy consumption is very great when the metro running. According to the actual statistics, the running energy consumption is about 590.6kW/km[1], and among which the ventilation and air conditioning system consumes about 40%. But in the winter, many waste heat produced by the communication and power transformation equipment must transfer to atmosphere by convection cooling air from fans, while the air temperature in staff area is always very low[2]. If the waste heat from the equipment rooms is recovered and transfer to the staff area, to improve the thermal comfort of staff, it is very significant for the metro energy saving.

In this paper, the thermal environment parameters were measured and analyzed, to evaluate the thermal comfort of staff area in winter, and to calculate the possible heat recover and to analyze the feasibility of waste heat transfer to staff area. In the winter of China north, like the coastal city of Qingdao, the cold climate is always longer, so the thermal environment and heat comforts are very important and necessary to be investigated, for the metro staff health and energy conservation. The piston wind effect is also considered in this paper, though the train's to and fro is in the lower layer, as shown in Fig.1 of waiting hall, the effect is obviously occurred in the Hall layer, and has influence to the staff working area.

2. Measurement of heat environment parameters

NO.3 line has been running now in Qingdao for one year, the subway station of WUSI Square and LICANG are the test sites during the coldest three months in this winter.

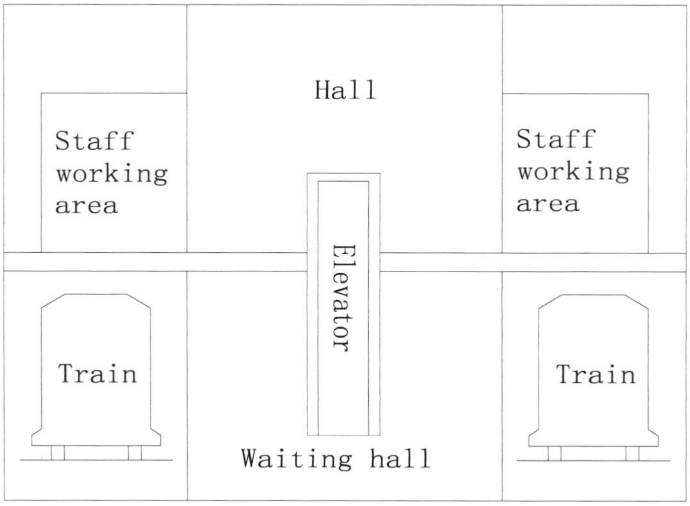

Figure 1. The metro station vertical view

2.1. Test sites and test aim

The measurement contents included temperature distribution, air flow velocity in Hall region; and the heat flux production in communication equipment room, comprehensive monitor room and transformation rooms; and then the length and width size of the Hall entrance and exit, tickets sale area, information area, security check area to calculate the heat load distance to arrive the heat comfort. There are wall-mounted sensors for the temperature and humidity, and its data are transfer to the monitor room. For the hall height is very higher than working staff, the heat comfort for staff should not judged by the data of monitor room. So the difference between the measurements with the wall-mounted sensor's data is also the gist to judge the real heat comfort and the necessary of improvement of thermal environment. So the measurement height is 1.5m, reflecting the staff real feeling.

2.2. Measuring apparatus

a. Temperature data acquisition instrument, AGILENT 34970A， the accuracy is 0.001℃.
b. Hot wire anemometer, SZX9-KA23, it can show the velocity and temperature and pressure. The accuracy of velocity test is 0.01m/s, and the accuracy of temperature test is 0.1℃.
c. Heat flux analyzer, the accuracy is ±2%.

3. Experimental results

3.1. Difference between the wall-mounted sensor with the measurement

The difference between the monitor data from wall-mounted sensor with the measurement data, is shown in Fig.2. The monitor data is higher than the measurement data for the wall-mounted sensor is nearer to the soil layer, and its temperature is the heat exchange result of soil with air flow in Hall. But the measurement data is just from the height of 1.5m distance to the ground. It is the convection heat exchange result of piston wind with the indoor air. The two temperature data are all increasing with the time, but the monitor data affect less influence from the piston wind, so it is not responding to the real feeling of the staff. The staff always feels the piston wind and the fluctuation temperature.

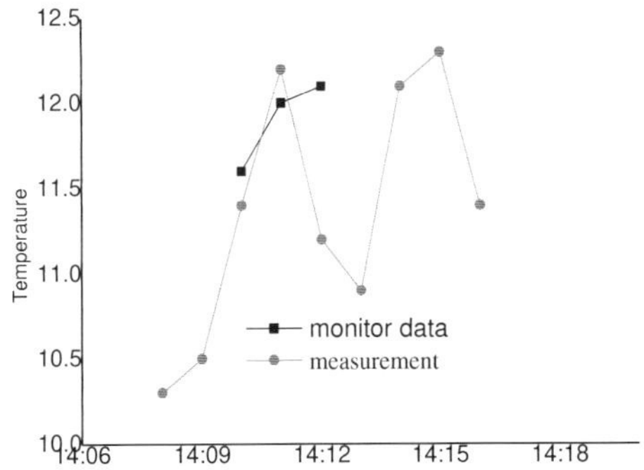

Figure 2. Difference between the monitor data with the measurement in public area in Feb.

3.2. Piston wind effect
At a security site, the detail measurement of the air temperature fluctuation with the piston wind[3], is shown as Fig.3.
When the piston wind is coming, the air temperature is decreasing, for the pumping cold air. The train drives the cold air in tunnel out to the waiting hall and then to Hall layer when it is coming to the station; and pump the colder air from the station entrance (outside) into the Hall when the train leaves the station. So the air temperature in the Hall is especially affected by the pumping when the train left and the entered colder air. One of the methods to improve the heat comfort could be preventing the pumping colder air entering the Hall.

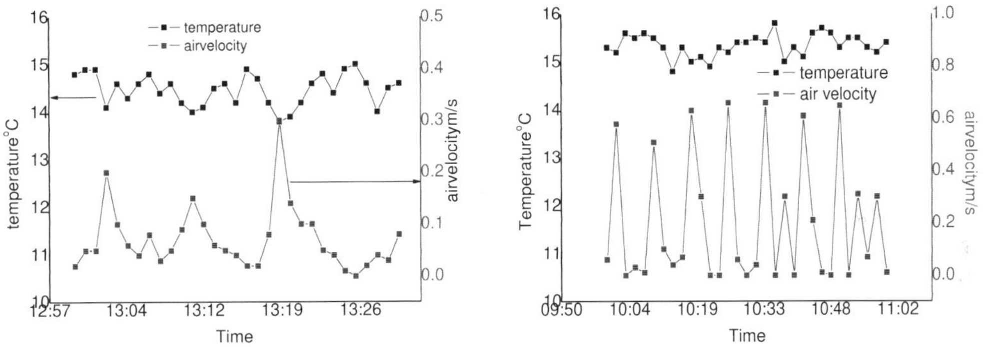

Figure 3. The piston effect on air temperature (left: March beginning; right: March end and windy day)

4. Heat load computation

From the measurement, the mean temperature of the staff working area in Hall in winter is just 12-13℃, whose is not satisfied with the heat comfort temperature of generally accepted 16℃ at least, so there need supplement heat. But as said in Introduction, the metro running energy consumption is so great that it is not accepted to supply heat to Hall by more energy consumption in winter. If the temperature of just at the staff working area not the whole Hall, rise to the 16 ℃, according the sizes of working are in WUSI station Hall, 44kW heat load should be supplied.

5. Feasibility analysis of the waste hear recovery to working area by heat pump

5.1. The heat flux measurement in equipment rooms
There are more than 10 rooms for varies equipment in WUSI station. The total mean heat flux produced by equipment is 38-120kW. For the room temperature arrives to 28-32℃, the cooling air fan

is stared and the 2.23 m^3/s air flow rate used to cool the equipment. The convection heat exchange makes the air outflow rise to 26 ℃ from 10℃ and then emitted to atmosphere. So the hot air flow could be the heat source for the waste heat recovery. The water-circulating heat pump could recover heat from the low temperature heat source, and transfer the more heat to the staff working area, including the consumed power convert to heat.

The co-efficient of performance of this heat pump could reach to 4.9 [4], so the input power is not large, 9 kW. The cooling water flowrate is 1.67 kg/s and the heating water flowrate is 1.05 kg/s. So the system transfer the waste heat to supply the heat to staff working area, which improved the heat comfort of metro staff in winter.

5.2. Introduction of the water-circulating heat pump system

The heat pump system is shown in Fig.4. The circulating water in heat pump absorb the heat Q1 from the return cooling air, and along with the input power to drive the compressor in heat pump, the Q2 heat supplement to the staff area to heat the indoor circulation air. Q2=Q1+P. The heat release ways would be determined in CFD[5-6] for optimization.

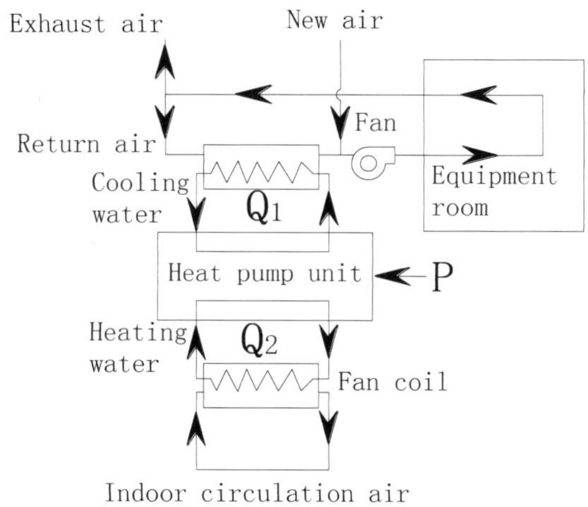

Figure 4. Water-circulating heat pump system to recover the waste heat

5.3. Economic analysis

The heat pump system adds two circulating pumps, one circulating air fan, and a heat pump unit. According to the power consumption (running cost) and the capital cost, the economy comparison could be drawn, compared with the traditional central heating method in China under the same heat comfort conditions.

According to the water and air circulating flowrate, the water pump of 1.5kW is chosen and the fan of 5kW is chosen. Along with the heat pump of 9kW consumption， the running time in winter is 873h, so the running power cost is about 5500RMB, less than 1000 $, it is superior to the central heating method, whose cost each winter is about 13000 RMB.

6. Conclusions

The measurement of temperature and air velocity in the metro staff working area in winter, show that the air temperature is not satisfied with the heat comfort though the air velocity fluctuation is not large by the piston wind effect. At the same time, the equipment (including the communication equipment, the power transformation equipment, and comprehensive control equipment) rooms release a large amount of heat and need to be cooled. So the waste heat could be transferred to the staff working area by the water-circulating heat pump to improve the heat comfort. The technique and economy of the waste heat recovery system are all feasible for the metro running energy conservation and thermal environment improvement.

7. References

[1] Hong Li Wang. Applied Mechanics and Materials, 2014, 3488-3641.

[2] Huan Zhang, Chunguang Zhu, Wandong Zheng, Shijun You, Tianzhen Ye, Peng Xue. Energy 2016, 116:880-893.

[3] George Katavoutas, Margarita N. Assimakopoulos, Dimosthenis N. Asimakopoulos. Science of the Total Environment 566–567 (2016):877–887.

[4] Yang Yao. China Building Industry Press, 2010.

[5] Yan Wu, Naiping Gao, Lihui Wang, Xiping Wu, Tong Zhu. Building Science (in Chinese), 2012, 28(8):70-76.

[6] Peigen Zhu, Yingxin Zhu, Xiaofeng Li. Underground Space (in Chinese), 2004, 24(2):161-165.

Acknowledgment

This work was supported by the National Natural Science Foundation of China (51676102) and the Natural Science Foundation of Shandong Province (ZR2015EM004). The authors greatly acknowledge the support from the Foundation of State Key Laboratory of Coal Clean Utilization and Ecological Chemical Engineering (Grant No. 2016-07).

IOP Conf. Series: Earth and Environmental Science **127** (2018) 012006 doi:10.1088/1755-1315/127/1/012006

Direct Solvothermal Liquefaction of *Pennisetum purpureum* Biomass to Produce Biocrude Using Ethanol Solvent

Mikael Adriane Ochiai, Salvador Marrod Cruz, Louiellyn Oporto, Rizalinda de Leon

Department of Chemical Engineering, University of the Philippines Diliman, Quezon City, Philippines

rizalinda.deleon@coe.upd.edu.ph, marrodcruz@gmail.com

Abstract. Direct solvothermal liquefaction was used in converting the lignocellulosic biomass, Pennisetum purpureum or Napier grass using ethanol as solvent. Liquefaction of Napier grass resulted in a dark and viscous bio-crude product and exhibited promising yields (34.6377% to 48.6267%). It was determined that the effects of temperature and residence time were statistically significant with the residence time having the greatest positive effect on yield. High yields of bio-crude from Napier grass seem to occur when solvothermal temperature, residence time increased and as solids ratio decreased. However, elemental analysis showed that the bio-crude produced needs to undergo deoxygenation (O: 14.25 – 49.42%) before mixing with petroleum. For the higher heating value (HHV), the parameters observed in the study were statistically insignificant, however, temperature was determined to have the greatest positive effect on HHV. It was observed that high HHV bio-crude were produced at high temperatures, low solids ratio, and low residence times. The acquired averages for the HHV (20.0333 MJ/kg to 29.7744 MJ/kg) were all higher than the HHV of the Napier grass sample used in the study (12.9394 MJ/kg). It was observed in the study, that even though solids ratio has the least effect on both responses, the choice of solids ratio is dependent on its interaction effects with the other parameters as its effects contribute to the observed responses.

1. Introduction

The The modern world is highly dependent on energy. About 96.7% of 9425 Mtoe of the total final consumption of the world by fuel in 2014 depends on oil, natural gas, coal, and other non-renewable energy sources and about 98.6% of 13699 Mtoe of the total primary energy supply by fuel in 2014 were non-renewable [1].

The continuous use of non-renewable energy sources may lead to a drastic decrease in supply resulting to different environmental issues. These led to researches on the discovery to use renewable energy sources including the use of chemical, geothermal, hydro, solar, and wind energy.

For the worldwide consumption of 15 TW of power for the year 2004, it was observed that solar energy has a high potential as its energy sources amount to 89000 TW. One means of extracting solar energy is through the indirect use of solar energy sinks such as lignocellulosic biomasses.

Bio-energy is the renewable energy stored in lignocellulosic biomasses in the form of chemical compounds, namely cellulose, hemicellulose, and lignin. Biomass most commonly refers to plant-derived organic matter. It was estimated to contribute about 10 – 14% of the primary energy supply of the globe [2].

One particular lignocellulosic biomass, Pennisetum purpureum, or commonly known as Napier grass, is a robust, reed-like perennial grass. This biomass has caught the attention of many researchers due to its unique characteristics. A specific study mentioned that Napier grass shows promise as a source for the acquirement of new products such as composites, cellulose derivatives, and biofuels [3].

In light of this, several methods to extract energy from the interconnected simple sugars present in cellulose and hemicellulose were developed [4]. Thermochemical conversion is the most used method in upgrading biomass to much more valuable products [5]. Hydrothermal liquefaction involves the use of water and possibly catalysts to directly liquefy pretreated lignocellulosic biomass resulting in the latter converting to liquid oil at temperatures below 400 °C [6].

However, due to the extreme operating conditions of hydrothermal liquefaction, other solvents such as ethanol may be used [7], [8]. Hence, solvothermal liquefaction was utilized in converting Napier grass as this opens up another potential energy source without compromising the environment since the use of renewable fuels from biomass provides for a carbon neutral energy source [7], [9], [4]. The effects of temperature, solids ratio, and residence time on the higher heating value (HHV), yield, and elemental components of the bio-crude were investigated in the study.

2. Materials and methodology

2.1. Preparation of biomass samples

The Napier grass samples collected from the Dairy Training and Research Institute in the University of the Philippines Los Baños were dried, milled, and sieved (Mesh 20-80). The particle size used in the study were the bottoms of the 80-mesh sieve (<0.177 mm). Water displacement method using a 10-mL graduated cylinder was performed to determine the density of the fine Napier grass samples used.

2.2. Effects of parameters

The study focused on three parameters: temperature, solids ratio, and residence time. Using a fractional factorial design of experiment, the low- and high-level of each factor is shown in Table I.

From this, necessary calculations for the amounts of biomass and solvent needed were performed based on the allowable volume of 4.5 mL of the reactor, and solids ratio, 10% and 20%, to be used in the study.

2.3. Solvothermal liquefaction

The liquefaction process was done in a sand bath heater with temperature control. The temperature of the reactor was allowed to reach the desired operating temperature, 230°C and 250 °C, before starting the countdown on the residence time, either 30 minutes or 60 minutes.

After the liquefaction process, the reactors were immediately quenched in running water. The product was extracted from the reactors using ethanol and was passed through a filter paper. The excess solvent was then allowed to evaporate before transferring into a vial. Further evaporation of the bio-crude was done until a constant mass was achieved.

2.4. Response and analysis

Once the bio-crude inside the vial achieved constant mass, the yield was calculated based on the amount of dried biomass used in the liquefaction process and the final mass of the bio-crude product.

For the HHV, around 0.2 g of each bio-crude product was added with ethylene glycol making sure that the total mass did not exceed 2 g. This was then subjected to bomb calorimetry in the Chemical Engineering Analytical Laboratory of the University of the Philippines – Diliman.

One sample for each run was sent for elemental analysis at the Department of Science and Technology to determine its hydrogen to carbon and oxygen to carbon ratios.

Statistical analysis was performed to determine the main effects with the aid of a statistical software, Minitab 17. The necessary statistics plots were also generated to determine the significance of the effects of each parameter on the response with a significance level α of 5%.

Table 1. Value of Low- and High-Level for each Factor

Factor	Variation Interval		
	Low Level	Basic Level	High Level
Temperature (°C), x1	230	240	250
Solids Ratio (%), x2	10	15	20
Residence Time (min), x3	30	45	60

3. Results and discussion

3.1. Biocrude synthesis
The study was successful in producing bio-crude by directly subjecting Napier grass to solvothermal liquefaction. Directly subjecting refers to liquefaction of the biomass without any chemical pretreatment performed. This was a concern in the study due to Napier grass having high lignin content, 30.40% [3] From this, it can be inferred that solvothermal liquefaction is capable of converting lignocellulosic biomass without pretreatment or catalyst. This also showed that Napier grass is indeed a promising lignocellulosic biomass for the production of different valuable materials as mentioned by [3].

3.2. Elemental analysis of biocrude and parameter effects
One way of determining the quality of bio-crude produced is through its elemental components. Results for the Elemental analysis of bio-crude produced along with a comparison of the typical ranges for the elemental components of petroleum are shown in Table II.

The % w/w of O was calculated by difference with the assumption that the % w/w of S is negligible. Determination of the elemental components of bio-crude is essential as bio-crude is used as an amendment to petroleum in refineries to address the demand for conventional fuels [10]. However, Table 2 shows that the elemental components of the crude produced does not fall between the ranges of the typical values for petroleum. This is the reason why direct use of bio-crude as transportation fuels prove to be a challenge, as it requires chemical modifications such as oxygen removal and molecular weight reduction to increase volatility, thermal stability and reduce viscosity [11].

It was observed in the study that increasing the levels of both temperature and solids ratio results in decreased atomic H/C ratio and that the inverse was true for the residence time. As with [11], an increase in temperature lead to an increase in the carbon content of the product resulting to a decrease in the atomic H/C ratio.

It was also observed through a Pareto Chart that the solids ratio had the greatest (0.2811) on the atomic H/C ratio, followed by temperature (0.2613), and then the residence time (0.2464). However, all parameters were insignificant to the response.

Table 2. Elemental Anaylysis of Bio-crude per run in % w/w Dry Basis

	Run 1.1	Run 2.1	Run 3.1	Run 4.1	Petroleum (Speight, 2006)
C	41.8	55.4	62.0	74.6	83.0 – 87.0
H	7.42	7.49	8.28	9.87	10.0 – 14.0
N	1.36	2.40	1.05	1.28	0.10 – 2.00
O	49.4	34.7	28.7	14.3	0.05 – 1.50
H/C	2.13	1.62	1.60	1.59	-
O/C	0.887	0.470	0.347	0.143	-

3.3. Higher heating value of biocrude produced and parameter effects

Another measure of the quality of bio-crude is its HHV. Table III shows the acquired HHV of each sample from bomb calorimetry.

It can be seen that the acquired average for the HHV ranges from around 20 to 30 MJ/kg, which is higher than the HHV of the Napier grass sample used, 12.9394 MJ/kg. This shows an increase of 54.82% for the lowest average, and an increase of 130.11% for the highest average. As a comparison, the acquired HHV in the study has a wider range (12.71 to 35.34 MJ/kg) compared to other studies: 28 to 34 MJ/kg [12], 27.9 to 33.8 MJ/kg [13], 32 to 34.7 MJ/kg [14], 35.9 to 39.6 MJ/kg [10], and 22.74 to 30.68 MJ/kg [11] due to the presence of an outlier.

It was observed in the study that an increase in the level of both temperature and solids ratio results in an increase in HHV, while the opposite was true for the residence time. This follows the work of [9] wherein it was mentioned that "an increase in temperature well above the activation energies for bond cessation causes extensive biomass depolymerization and that increasing the solids ratio allows for more sources of hydrogen and carbon for the final crude product". For the residence time, this may be explained by the fact that since Napier grass has a high oxygen content, allowing it to fully degrade might result to the final crude product having more oxygen, thus a higher atomic O/C ratio, resulting in low HHV.

The parameter that had the highest effect according to the Pareto Chart was temperature (1.3976), followed by residence time (1.2882), and then the solids ratio (0.5955). However, all parameters were statistically insignificant to the HHV.

3.4. Yield of biocrude and parameter effects

Another objective of the study is to be able to produce bio-crude with high yields. Table IV shows the yields of bio-crude in the study.

The acquired average for the yield ranges from around 34 to 49%. As a comparison, the yield acquired from the study is the highest (34.64 to 48.63%) when compared to other studies: 3 to 13% [12], ~10 to 15% [13], ~2.5 to 37.5% [14], 9.4 to 32.6% [10], and 21.8 to 36.5% [11]. This enforces the idea that solvothermal liquefaction is capable of producing higher yields of bio-crude compared to hydrothermal liquefaction. However, an appropriate comparison should be done for hydrothermal liquefaction of Napier grass with the same conditions as in the study.

Table 3. Elemental Anaylysis of Bio-crude per run in % w/w Dry Basis

Run	y_{u1}	y_{u2}	y_{u3}	$\bar{y}_u$	σ
1	18.7686	28.6193	12.7121	20.0333	8.0287
2	20.1542	28.3525	32.0887	26.8651	6.1047
3	28.9203	27.7182	32.6846	29.7744	2.5910
4	35.3421	23.8814	22.5624	27.2620	7.0286

Reference [15] observed that the kinetics of esterification of fatty acids in alcohols at critical supercritical conditions were influenced by numerous factors. In the study performed, it was observed that an increase in the response of yield occurs when the level of both temperature and residence time increases, while the opposite is true for the solids ratio. This shows that the yield of bio-crude from solvothermal liquefaction is affected by the physicochemical properties of the alcohol at different operating conditions as with the esterification of fatty acids.

It was observed that higher values of biomass to solvent ratio results in increased formation of solid residues as with [2]. This may have resulted from the decrease in relative interactions between the molecules of biomass and that of ethanol at high biomass to solvent ratios resulting to the suppression of the dissolution of biomass components as explained by [9]. It was also observed in the study that the yield of bio-crude decreased as the solids ratio was decreased. This was also observed by [16]. Thus, it can be concluded that the increase in solids residue and decrease in bio-crude oils are relatively based on the solids ratio, wherein the solids residue and bio-crude produced can be thought of as a percentage of the solids ratio. It was observed that yield of bio-crude from Napier grass seems to increase as solvothermal temperature, residence time increased and as solids ratio decreased.

It was determined using a Pareto Chart that the residence time has the greatest effect (4.3661) on yield, followed by temperature (2.4604), and then the solids ratio (2.2461). Additionally, the effects of residence time and solids ratio were statistically significant to the yield of bio-crude. This is supported by a study performed by Akhtar and Amin [9] wherein they mentioned that temperature has a synergetic effect on the yield of bio-crude and that the residence time may dictate the overall conversion of biomass.

Table 4. Table of Yield for Each Sample in %

Run	y_{u1}	y_{u2}	y_{u3}	$\bar{y}_u$	σ
1	43.4129	48.9832	52.1662	48.1874	4.4306
2	33.6439	34.0373	36.2319	34.6377	1.3946
3	41.6446	41.9098	49.2927	44.2824	4.3411
4	51.0352	45.0518	49.7930	48.6267	3.1576

4. Conclusion

In the study, an energy efficient way for producing bio-crude oil from lignocellulosic biomass, Pennisetum purpureum or Napier grass, by direct solvothermal liquefaction using ethanol as solvent was done and the effects of various parameters on the yield and quality of the biocrude produced was explored.

Elemental analysis shows that the bio-crude produced do not fall the in typical range of components of petroleum, thus renders the product immiscible with the latter. The oxygen content for all products are considerably large, thus a need for deoxygenation arises. This high oxygen content may be explained by the inefficient liquefaction of the biomass used or that equilibrium conversion was not yet achieved.

The acquired averages for the HHV of the bio-crude produced were all higher than the HHV of the Napier grass sample used in the study. It was determined that temperature has the greatest effect on the HHV. Increasing the residence time has a negative effect on the HHV of the bio-crude. However, it was observed that the effects of the parameters were statistically insignificant to the HHV. For yield, however, temperature and residence time were statistically significant. It was observed that residence time has the greatest effect on yield. However, increasing the solids ratio is shown to have a negative contribution on the yield of bio-crude.

It was observed that despite the solids ratio having the least effect on both responses, the choice of solids ratio greatly depends on its interaction with the other parameters observed in the study.

The The modern world is highly dependent on energy. About 96.7% of 9425 Mtoe of the total final consumption of the world by fuel in 2014 depends on oil, natural gas, coal, and other non-renewable energy sources and about 98.6% of 13699 Mtoe of the total primary energy supply by fuel in 2014 were non-renewable [1].

5. References

[1] IEA (September 2016) Key World Energy Statistics. Paris, France.

[2] C. Wang, J. Pan, J. Li, and Z. Yang, "Comparative studies of products produced from four different biomass samples via deoxy-liquefaction," *Bioresource Technology*, pp. 2778-2786, 2008.

[3] A.D. Morandim-Giannetti, T.S. Albuquerque, R.K. de Carvalho, R.M. Araújo, and R. Magnabosco, "Study of napier grass delignification for production of cellulosic derivatives,' *Carbohydrate Polymers*, pp. 849-855, 2013.

[4] E.-M. Aro, "From first generation biofuels to advanced solar biofuels," Ambio, pp. 24-31, 2016.

[5] C. Zhong and X. Wei, "A comparative experimental study on the liquefaction of wood," Energy, pp. 1731-1741, 2004.

[6] Y. Zhang, "Hydrothermal liquefaction to convert biomass into crude oil," in Biofuels from

Agricultural Wastes and Byproducts, H. Blaschek, T. Ezeji, & J. Scheffran, Eds. Hoboken: Blackwell Publishing, 2010, pp. 201-232.

[7] C. Xu and T. Etcheverry, "Hydro-liquefaction of woody biomass in sub- and super-critical ethanol with iron-based catalysts," Fuel, 335-345, 2008.

[8] J. Zhang, Z. Luo, and Y. Zhang, "Hydrothermal liquefaction of Chlorella pyrenoidosa in water and ethanol," *Energy Procedia*, pp. 253-259, 2013.

[9] J. Akhtar and N.A. Amin, "A review on process conditions for optimum bio-oil yield in hydrothermal liquefaction of biomass," *Renewable & Sustainable Energy Reviews*, pp. 1615-1624, 2011.

[10] J.M. Jarvis, J.M. Billing, R.T. Hallen, A.J. Schmidt, and T.M. Schaub, "Hydrothermal liquefaction biocrude compositions compared to petroleum crude and shale oil," *Energy & Fuels*, pp. 1-47, 2017.

[11] D. Zhou, S. Zhang, H. Fu, H., and J. Chen, "Liquefaction of macroalgae Enteromorpha prolifera in sub-/supercritical alcohols: direct production of ester compounds," *Energy & Fuels*, p. 2342, 2012.

[12] H. Ramsurn, "Gasification, liquefaction and deoxy-liquefaction of switchgrass using sub- and supercritical water," Auburn, Alabama, United States of America, August 2013.

[13] C. Jazrawi, P. Bille, A.B. Ross, A. Montoya, T. Maschmeyer, and B.S. Haynes, "Pilot plant testing of continuous hydrothermal liquefaction of microalgae," Algal Research, pp. 268-277, 2013.

[14] D.R. Vardon et al., "Chemical properties of biocrude oil from the hydrothermal liquefaction of Spirulina algae, swine manure, and digested anaerobic sludge," *Bioresource Technology*, pp. 8295-8303, 2011.

[15] S. Changi, T. Pinnarat, and P.E. Savage, "Modeling hydrolysis and esterification kinetics for biofuel processes," *Industrial & Engineering Chemistry Research*, pp. 3206-3211, 2011.

[16] D.G. Boocock and K.M. Sherman, "Further aspects of powdered poplar wood liquefaction by aqueous pyrolysis," *The Canadian Journal of Chemical Engineering*, pp. 627-633, 1985.

Acknowledgment
The authors gratefully acknowledge the support of the Alma Bella P. Madrazo Thesis Support Grant through the UP Engineering Research and Development Foundation, Inc.

A logistic regression approach to model the willingness of consumers to adopt renewable energy sources

M M Ulkhaq[1], A K Widodo[1], M F A Yulianto[1], Widhiyaningrum[1], A Mustikasari[1], P Y Akshinta[1]

[1] Department of Industrial Engineering, Diponegoro University, Semarang, Indonesia

ulkhaq@live.undip.ac.id

Abstract. The implementation of renewable energy in this globalization era is inevitable since the non-renewable energy leads to climate change and global warming; hence, it does harm the environment and human life. However, in the developing countries, such as Indonesia, the implementation of the renewable energy sources does face technical and social problems. For the latter, renewable energy sources implementation is only effective if the public is aware of its benefits. This research tried to identify the determinants that influence consumers' intention in adopting renewable energy sources. In addition, this research also tried to predict the consumers who are willing to apply the renewable energy sources in their houses using a logistic regression approach. A case study was conducted in Semarang, Indonesia. The result showed that only eight variables (from fifteen) that are significant statistically, i.e., educational background, employment status, income per month, average electricity cost per month, certainty about the efficiency of renewable energy project, relatives' influence to adopt the renewable energy sources, energy tax deduction, and the condition of the price of the non-renewable energy sources. The finding of this study could be used as a basis for the government to set up a policy towards an implementation of the renewable energy sources.

1. Introduction

All forms of energy sources have negative impacts on sustainable development values. Fossil fuels, such as coal, oil, and natural gas, do substantially more harm by most measures, including air and water pollution, damage to public health, wildlife and habitat loss, as well as water and land use. Consumption of fossil fuels resources, which were formed from prehistoric plants and animals that live hundreds of millions of years ago, leads to global warming and climate change. Although the fossil fuels do harm the environment and human life, their global consumption increased of about 0.6% in 2016 [1]. As a result, global carbon dioxide emissions set a new all-time record high! Carbon dioxide emissions in 2015 were 36 million metric tons higher than in 2014 and marked the sixth straight year a new record high has been set. This situation actually cannot be allowed to continue in the future. Since the fossil fuels come from buried dead organisms, they do have a limited amount—it is non-renewable. They are expected to run out 30 to 60 more years [2]; hence, it is only a matter of time when they are diminished!

In Indonesia, as a developing country, the consumption of fossil fuel energy sources to the total energy sources is increasing year-by-year. In 2014, the consumption of fossil fuel absorbs more than 65% of the total energy sources; in contrast with Sweden, Finland, and France with less than 50% of the total energy consumption [3]. Generally, the fossil fuels in Indonesia are used the most in the industrial sector, which is about 296 million tons of oil equivalent (Mtoe), or about 41% from total consumption of national energy; followed closely by transportation sector which consumes 226.6 Mtoe (37%) in 2009 [4].

This condition makes the government and the energy society worry as the fossil fuels energy resources will be diminished in the future. In addition, the increasing dependency on the non-renewable energy sources could influence negative impact toward the environment and become a trigger to global warming. Changing the energy sources from non-renewable to renewable is necessary. Several countries have changed their energy usage from non-renewable to renewable; for instance, the U.S. Environmental Protection Agency in 2014 has proposed a clean power plan, which intends to diminish U.S. carbon pollution by 30% in 2030 [5].

Renewable energy comes from regenerative resources that cannot be depleted, such as solar, hydropower, wind, biomass, and geothermal. It is sustainable, ubiquitous, and essentially non-pollute. It has various advantages, such as strengthening the economy to reduce the reliance on fossil fuels and making the nations more secure in providing pollution-free energy [6]–[8]. As the impacts of climate change worsen with each passing year, replacing fossil fuels with renewable energy will be imperative to the protection of wildlife and wild lands.

Fortunately, the development of the renewable energy sources is growing remarkably from year to year. It includes the continuation of comparatively low global fossil fuel prices, dramatic price declines of several renewable energy technologies, and a continued increase in attention to energy storage. As of 2015, renewable energy provided an estimated 19.3% of global final energy consumption, and growth in capacity and production continued in 2016. In addition, the renewable energy sector employed 9.8 million people in 2016, an increase of 1.1% over 2015 [9].

However, the utilization of renewable energy resources in Indonesia has not been optimized due to its high production cost and it is worsened by the subsidy policy on fossil energy. Moreover, some opposite perceptions from the citizen about the renewable energy make it more difficult to implement it. Public perceptions, consciousness, and acceptance of renewable energy are significant social factors, which shall be taken into consideration in establishing prospective energy systems. Implementation and deployment of the renewable energy can be sustainable and effective only in the case if the public is aware of its benefits and its needs in the deployment of the renewable energy on the human activities and on the environment [10].

Several previous researches have been conducted to address consumers' stated intentions to adopt the renewable energy. According to [11]–[13], consumers with higher income are willing to adopt renewable energy. Age had a significant impact on influencing the willingness to adopt the renewable energy [14], yet [11] found it conversely. Other researchers found that higher education is the main individual characteristics in adopting the renewable energy [15]. However, these outcomes are based on the results of individual researches, and may not be conclusive across studies so as errors will appear when transferring estimates from one site to another.

Therefore, in this study, we focus on investigating the determinants that influence consumers' intention towards the adoption of renewable energy sources in the household sector. A case study was conducted in Semarang, Indonesia, to model the willingness of the consumers to adopt the renewable energy sources. The identification of the characteristic of the potential market segment is becoming a major reason in renewable energies implementation in Indonesia.

2. Research methods

A logistic regression was employed in this research to model the willingness of consumers to adopt renewable energies. It is frequently used to explain a model that connects one non-metric dependent variable and several—metric and/or non-metric—independent variables. Its primary objectives are to understand group differences and to predict the likelihood that an entity (individual or object) will belong to a particular class or group based on those several metric independent variables [9], [16].

A logistic regression is similar to multiple regression analysis in that one or more independent variables are used to predict a single dependent variable. What distinguishes a logistic regression model from multiple regression is that the dependent variable is nonmetric. It is preferable to discriminant analysis since the former can accommodate all types of independent variables (metric and nonmetric) and do not require the assumption of multivariate normality [17].

Since the objective of this research is to investigate the consumers' attitude towards the adoption of renewable energy sources, we set the dependent variable (Y) as the willingness of the consumers to adopt renewable energies. The score of 1 (one) was given when the consumer *i* is willing to adopt the renewable energy sources in his/her house; while the score of 0 (zero) was given to the opposite answer. For the set of independent variables (X_j), we follow the study by [18] as they studied the determinants that affect consumers' intention towards the adoption of renewable energy sources in the residential sector in Greece. The model is then as follows:

$$Y = \beta_0 + \beta_1 X_1 + \beta_2 X_2 + \beta_3 X_3 + \beta_4 X_4 + \beta_5 X_5 + \beta_6 X_6 + \beta_7 X_7 + \beta_8 X_8 + \beta_9 X_9 + \beta_{10} X_{10} + \beta_{11} X_{11} + \beta_{12} X_{12} + \beta_{13} X_{13} + \beta_{14} X_{14} + \beta_{15} X_{15} + \varepsilon, \quad (1)$$

where β_0 is a constant; β_j is the regression coefficient ($j = 1 \ldots 15$); X_1 is a dummy variable for gender (1 if the consumer is female and 0 if the consumer is male); X_2 is a metric variable for the consumer's age; X_3 is a metric variable for the square of the consumer's age; X_4 is a dummy variable for the marital status of the consumer (1 if the consumer is married and 0 otherwise); X_5 is a dummy variable for the educational background of the consumer (1 if the consumer has completed undergraduate studies and 0 otherwise); X_6 is a dummy variable for the employment status of the consumer (1 if the consumer is a public servant employee and 0 otherwise); X_7 is a metric variable for the consumer's monthly private income in Rupiah (Indonesian currency); X_8 is a dummy variable for the consumer's status of home ownership (1 if the consumer owns his/her house and 0 otherwise); X_9 is a metric variable expressing the average electricity cost per month in Rupiah; X_{10} is a dummy variable taking the value 1 if the consumer considers that the installation and maintenance costs of renewable energies in the residential sector are too high and 0 otherwise; X_{11} is a dummy variable accounting for 0 if the consumer is uncertain about the efficiency of domestic renewable energy projects and 1 otherwise; X_{12} is a dummy variable accounting for 1 if the consumer would install a renewable energy system in his/her house if his/her family or friends would encourage him/her to do so and 0 otherwise; X_{13} is a dummy variable taking the value 1 if the consumer considers that an energy tax deduction would encourage him/her to adopt renewable energy sources in his/her house and 0 otherwise; X_{14} is a dummy variable accounting for 1 if the consumer would adopt renewable energy sources in his/her house in case of an energy subsidy and 0 otherwise; X_{15} is a dummy variable taking the value 1 if the consumer is willing to adopt renewable energy projects in his/her home as a result of energy prices of non-renewable energy sources doubling and 0 otherwise; and ε is an error term.

The value of $b_0, b_1, \ldots, b_{15}$, as unbiased estimators of β*'s* can be found by employing the method of maximum likelihood. The likelihood L is given by the joint probability distribution evaluated at the observed counts Y_i; hence:

$$L(b_0, b_1, \ldots b_{15}) = \frac{\prod_{i=1}^{n} e^{y_i(b_0 + b_1 X_{i1} + \ldots + b_{15} z_{i15})}}{\prod_{i=1}^{n} \left(1 + e^{b_0 + b_1 X_{i1} + \ldots + b_{15} z_{i15}}\right)}. \quad (2)$$

The values of the coefficients that maximize the likelihood cannot be expressed in a nice closed form solution as in the normal theory linear models case. Instead, they must be determined numerically by starting with an initial guess and iterating to the maximum of the likelihood function. Technically, this procedure is called an iteratively reweighted least squares method (see [19] for the detail).

The predicted probabilities of the consumer belong to the certain group, i.e., 1 or 0 can be calculated using the formula for the cumulative distribution function of the standard logistic distribution. For example, the predicted probabilities of the consumer belong to the group 1, i.e., he/she is willing to adopt the renewable energy sources in his/her residence is:

$$\Pr(Y = 1) = \frac{e^{(b_0 + b_1 X_1 + \ldots + b_{15} X_{15})}}{1 + e^{(b_0 + b_1 X_1 + \ldots + b_{15} X_{15})}}. \quad (3)$$

3. Case study

A case study was conducted in Semarang, Indonesia, to investigate the resident's willingness to apply renewable energy sources in his/her house. The requirements to be participants in this research are over 16 years old and have any specific jobs or they have their own incomes. The potential participants were first approached and asked if they agreed to participate in the survey. Two hundred respondents participated in the survey. Profile of the respondents is shown in Table 1.

Table 1. Profile of the respondents.

Variable	Percentage
Sex:	
Male	53.0
Female	47.0
Age in years:	
< 21	2.5
21–30	21.5
31–40	30.5
41–50	26.0
> 50	19.5
Occupation:	
Public servant employee	43.0
Others	57.0
Educational level:	
Undergraduate level	55.5
Graduate level	12.0
Others	32.5

To investigate consumers' awareness towards renewable energy sources, we questioned if they knew that it is possible to produce energy for household use from the implementation of five alternative renewable energy sources, i.e., biomass, geothermal, hydropower, solar, and wind. About 88% of the respondents were aware of the solar energy, about 76.5% were aware of the hydropower energy, 68% of the wind energy, 52.5% of the biomass, and geothermal energy was the least well-known (48.5%). This result is specifically depicted in Figure 1. Note that from 200 respondents, none of them had previously used renewable energy sources for their household's electricity supply.

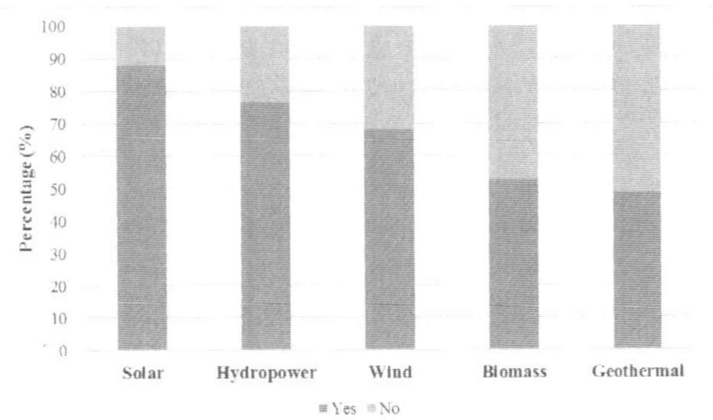

Figure 1. Consumers' awareness of the potential use of renewable energy technologies.

A logistic regression was then employed to estimate the model in (1). Table 2 summarizes the empirical result the logit equation's estimated coefficients with respect to consumers' intention in Semarang to adopt renewable energy sources in their houses. Note that only eight independent variables are statistically significant: X_6, X_9, and X_{15} are at the levels of 10%; X_5, X_7, X_{12}, and X_{13} are at the level of 5%, and only X_{11} is significant at the level of 1%. The model gives the Nagelkerke R^2 [20]

of 0.519 which is considered as high. It means that the independent variables could explain 51.9% of the total (100%) variability of the dependent variables. Next, the Hosmer and Lemeshow test [21], [22] shows that the model is significant, i.e., the model has significance value of 0.926, which is above the significance level of 5%.

Table 2. Logistic regression result.

Independent Variables	b	exp(b)	Standard of Error	Significance
Constant	–5.198	0.006	3.497	0.138
X_1	–0.651	0.522	0.445	0.144
X_2	–0.023	0.977	0.168	0.890
X_3	0.000	1.000	0.002	0.854
X_4	–0.601	0.548	0.783	0.443
X_5	–1.288	0.276	0.565	0.023[b]
X_6	–0.719	0.487	0.435	0.098[c]
X_7	0.000	1.000	0.000	0.010[b]
X_8	–0.322	0.725	0.699	0.645
X_9	0.000	1.000	0.000	0.063[c]
X_{10}	0.387	1.472	0.470	0.410
X_{11}	1.694	5.443	0.470	0.000[a]
X_{12}	1.553	4.725	0.600	0.010[b]
X_{13}	2.522	12.452	1.098	0.022[b]
X_{14}	0.163	1.177	0.806	0.839
X_{15}	0.937	2.554	0.560	0.094[c]

[a] Levels of significance at 1%.
[b] Levels of significance at 5%.
[c] Levels of significance at 10%.

Educational background of the respondents (X_5) is statistically significant according to Table 2. This result indicates that educated people are more willing to implement the renewable energy sources in their house; vice versa (in line with the study by [15], [18]). However, X_1, X_2, X_3, and X_4 are not statistically significant. It seems that the willingness of the respondents to adopt the renewable energy sources is not affected by gender, age, and marital status. We cannot say that male people tend to be more willing to adopt renewable energy sources than female people are, or younger consumers are less willing to pay extra than middle-aged consumers are (in contrast with the study by [14] which stated that age is significant), or unmarried people are less willing to implement renewable energy sources in their house.

We also empirically examined the effect of financial measures on the consumers' intention to adopt renewable energy sources in their house. The result is not surprising since most of "financial" independent variables are statistically significant, i.e., X_6, X_7, X_9, X_{13}, and X_{15}. Public servant employees, high-income people, or people who pay more on their electricity costs tend to be more willing to install renewable energy sources in their residence (along with the study by [11]–[13]). Tax deduction policy or increasing of the price of non-renewable energy sources also have the effect to influence people to adopt renewable energy sources. Interestingly, the respondents' status of home ownership does not affect their attention towards renewable energy sources.

The rest of the variables are X_{10} (not significant), X_{11} (significant), X_{12} (significant), and X_{14} (not significant). It means that we cannot conclude that the consumers would adopt the renewable energy sources if they think that the installation and maintenance costs are too high. Energy subsidy, however, has no effect to the decision of the consumers to apply the renewable energy into their house. On the other hand, the assurance of the efficiency to apply the renewable energy sources gives significance impact on the decisions of the consumers. It means that if the renewable energy project cannot guarantee that the project would give better usage efficiency than fossil fuels, the consumers would not apply it. Consumers' decisions also are affected by the influence of their relatives. In other words, the consumers would install the renewable energy system in their houses if their families or friends would encourage them to do.

The logistics regression approach is not only able to identify the variables that affect the willingness of the consumers to adopt the renewable energy sources (see Table 2 for the result), but also it can predict the behavior of the "other consumers" based on their characteristics. The formula in (3) can be used to employ the task. An example is shown as follows.

A 25 years old unmarried male consumer would be identified if he were willing to adopt the renewable energy sources into his own house. He has been graduated from industrial engineering department (bachelor degree) and now is a public servant employee in Semarang. He could earn 5 million rupiahs per month as he pays two hundred thousand for electricity. He stated that he is sure about the efficiency of the renewable energy project since he has read many kinds of literature about the benefits gained from the renewable energy sources. In fact, he does not worry about whether the government would give a tax deduction or not; or give an energy subsidy or not. However, he does concern if the price of non-renewable energy source would be doubled. Since he is a self-confident person, he is not affected by the influence of his relatives in his decision to adopt the renewable energy sources. Based on the scenario, we could discover the value of each variable as follows. The value of X_1 is 1, X_2 is 25, X_3 is 625, X_4 is 0, X_5 is 1, X_6 is 1, X_7 is 5,000,000, X_8 is 1, X_9 is 200,000, X_{10} is 0, X_{11} is 1, X_{12} is 0, X_{13} is 0, X_{14} is 0, and X_{15} is 1. From (3), the probability of the consumer would adopt the renewable energy sources into his house is:

$$\Pr(Y=1)=\frac{e^{(b_5X_5+b_6X_6+b_7X_7+b_9X_9+b_{11}X_{11}+b_{12}X_{12}+b_{13}X_{13}+b_{15}X_{15})}}{1+e^{(b_5X_5+b_6X_6+b_7X_7+b_9X_9+b_{11}X_{11}+b_{12}X_{12}+b_{13}X_{13}+b_{15}X_{15})}}$$

$$\Pr(Y=1)=\frac{e^{(1(-1.288)+1(-0.719)+0+0+1(1.694)+0+0+1(0.937))}}{1+e^{(1(-1.288)+1(-0.719)+0+0+1(1.694)+0+0+1(0.937))}}$$

$$\Pr(Y=1)=0.651.$$

Note that only X_5, X_6, X_7, X_9, X_{11}, X_{12}, X_{13}, and X_{15} are significant. Since the probability is 0.651 or above the cut-off point of 0.5, we could conclude that the consumer is willing to adopt the renewable energy sources in his house.

4. Conclusion and future research direction

The objectives of this study are twofold. First is to identify the determinants that influence consumers' intention towards the adoption of renewable energy sources in the household sector. The second is to model and predict the consumers' willingness to adopt the renewable energy sources in their houses. A case study to exhibit the applicability of this study has been conducted in Semarang, a capital city of Central Java Province, Indonesia.

The result of the case study showed that from fifteen variables that are employed, only eight variables are statistically significant. There are educational background, employment status, consumers' incomes per month, consumers' average electricity costs per month, certainty about the efficiency of renewable energy project, relatives' (families or friends) influence to adopt the renewable energy sources, energy tax deduction, and the condition of the price of the non-renewable energy sources (the complete result can be seen in Table 2). It can be concluded that that educated people, public servant employees, high-income people, people who pay more on their electricity costs, people who are sure about the efficiency of the renewable energy project, people whose decisions are easily affected by their relatives' decisions, people who concern of energy tax deduction, and people who worry if the price of the non-renewable energy sources would double, are more willing to adopt the renewable energy sources.

The finding of this study could be used as a basis for the government to set up a policy toward renewable energy implementation. A system dynamics approach—initially developed by [23]—can be utilized to formulate the policy. It is frequently used when facing high-stakes decisions and seeking an integrated view of the major forces that can affect key outcomes years or decades into the future [24].

It has been used in formulating the policies related to energy; see for example [25]–[27]. Using the system dynamics approach to support the government in preparing the policy about renewable energy is an interesting area to be pursued.

5. References

[1] BP 2016 *BP Statistical Review of World Energy*
[2] Shafiee S and Topal E 2008 When fossil fuel reserves be diminished? *Energy Policy* vol 37 pp 181–189
[3] World Bank 2014 *Fossil Fuel Energy Consumption (% total)*
[4] Hasan M H, Mahlia T M I and Nur H 2012 A review on energy scenario and sustainable energy in Indonesia *Renewable and Sustainable Energy Reviews* vol 16 pp 2316–2328
[5] Soon J J and Ahmad S A 2015 Willingly or grudgingly? A meta-analysis on the willingness-to-pay for renewable energy use Renewable and Sustainable Energy Reviews vol 44 pp 877–887
[6] Lee C Y and Heo H 2016 Estimating willingness to pay for renewable energy in South Korea using the contingent valuation method *Energy Policy* vol 94 pp 150–156
[7] Ma C, Rogers A A, Kragt M E, Zhang F, Polyakov M, Gibson F, Chalak M, Pandit R and Tapsuwan S 2015 Consumers' willingness to pay for renewable energy: A meta-regression analysis, *Resource and Energy Economics* vol 42 pp 93–109
[8] Murakami K, Ida T, Tanaka M and Friedman L 2015 Consumers' willingness to pay for renewable and nuclear energy: a comparative analysis between the US and Japan *Energy Economics* vol 50 pp 178–189
[9] REN21 2017 *Renewables 2017 Global Status Report*
[10] Yazdanpanah M, Komendantova N and Ardestani R S 2015 Governance of energy transition in Iran: Investigating public acceptance and willingness to use renewable energy sources through socio-psychological model *Renewable and Sustainable Energy Reviews* vol 45 pp 565–573
[11] Zoric J and Hrovatin N 2012 Household willingness to pay for green electricity in Slovenia *Energy Policy* vol 47 pp 180–187
[12] Yoo S H and Kwak S Y 2009 Willingness to pay for green electricity in Korea: a contingent valuation study *Energy Policy* vol 37 pp 5408–5416
[13] Menges R, Schroeder C and Traub S 2005 Altruism, warm glow and the willingness-to-donate for green electricity: an artefactual field experiment *Environmental and Resource Economics* vol 31 pp 431–458
[14] Hast A, Alimohammadisagvand B and Syri S 2015 Consumer attitudes towards renewable energy in China—the case of Shanghai *Sustainable Cities and Society* vol 17 pp 69–79
[15] MacPherson R and Lange I 2012 Determinants of the green electricity tariff uptake in the UK *Stirling Economics Discussion Paper*, vol 2012-11 pp 1–28
[16] Johnson R A and Wichern D W 2007 *Applied Multivariate Statistical Analysis* vol 6 (Upper Saddle River: Pearson Prentice Hall)
[17] Hair J F, Black W C, Babin B J and Anderson RE 2009 *Multivariate Data Analysis* vol 7 (Upper Saddle River: Pearson Prentice Hall)
[18] Sardianou E and Genoudi P 2013 Which factors affect the willingness of consumers to adopt renewable energies? *Renewable Energy* vol 57 pp 1–4
[19] McCullagh P and Nelder J A 1989 *Generalized Linear Models* vol 2 (London: Chapman and Hall)
[20] Nagelkerke N J D 1991 A note on a general definition of the coefficient of determination *Biometrika* vol 78 pp 691–692
[21] Hosmer D W and Lemeshow S 1980 A goodness-of-fit test for the multiple logistic regression model Communications in Statistics vol A10 pp 1043–1069.
[22] Lemeshow S and Hosmer D W 1982 The use of goodness-of-fit statistics in the development of logistic regression models *American Journal of Epidemiology* vol 115 pp 92–106
[23] Forrester J W 1961 *Industrial Dynamics* (Cambridge: The MIT Press)
[24] Homer J B 2012 *Models That Matter: Selected Writings on System Dynamics 1985-2010* (Grapeseed Press)

IOP Conf. Series: Earth and Environmental Science **127** (2018) 012007 doi:10.1088/1755-1315/127/1/012007

[25] Aslani A, Helo P and Naaranoja M 2014 Role of renewable energy policies in energy dependency in Finland: system dynamics approach Applied Energy vol 113 pp 758–765
[26] Feng Y Y, Chen S Q and Zhang L X 2013 System dynamics modeling for urban energy consumption and CO_2 emissions: a case study of Beijing, China *Ecological Modelling* vol 252 pp 44–52
[27] Hsu C W 2012 Using a system dynamics model to assess the effects of capital subsidies and feed-in tariffs on solar PV installations Applied Energy vol 100 pp. 205–217

IOP Conf. Series: Earth and Environmental Science **127** (2018) 012008 doi:10.1088/1755-1315/127/1/012008

Analysis on Behaviour of Wavelet Coefficient during Fault Occurrence in Transformer

Bancha Sreewirote[1], Atthapol Ngaopitakkul[2]

[1] Department of Electrical Engineering, Faculty of Engineering Thonburi University, Bangkok, Thailand
[2] Faculty of Engineering, King Mongkut's Institute of Technology Ladkrabang, Bangkok, Thailand

atthapol.ng@kmitl.ac.th

Abstract. The protection system for transformer has play significant role in avoiding severe damage to equipment when disturbance occur and ensure overall system reliability. One of the methodology that widely used in protection scheme and algorithm is discrete wavelet transform. However, characteristic of coefficient under fault condition must be analyzed to ensure its effectiveness. So, this paper proposed study and analysis on wavelet coefficient characteristic when fault occur in transformer in both high- and low-frequency component from discrete wavelet transform. The effect of internal and external fault on wavelet coefficient of both fault and normal phase has been taken into consideration. The fault signal has been simulate using transmission connected to transformer experimental setup on laboratory level that modelled after actual system. The result in term of wavelet coefficient shown a clearly differentiate between wavelet characteristic in both high and low frequency component that can be used to further design and improve detection and classification algorithm that based on discrete wavelet transform methodology in the future.

1. Introduction

Electrical power system has been constantly expanding and getting more complex to keep up with consumption demand due to economic and population growth. It mainly consists of generation, transmission, and distribution system. One of the most vital equipment in power system due is transformer due to its function that can transfer power from one voltage level to another with little power loss in process. This role of transformer enables long distance power transmission from power plant to end-user via transmission line. To ensure reliability of the system, the protection system especially for transformer must be response to disturbance such as fault in system with accurate and fast in order to provide correct response for operator. This can prevent severe damage to transformer and avoid economic cost that result from cost of equipment, operating cost, and loss of opportunity during replacement of damaged transformer.

From the important issue mentioned above algorithm for detect, located and classify disturbance in transformer has been constantly develop and improve. One of the methodology has been widely applied on power system protection is Discrete Wavelet Transform (DWT). The reason is its main attribute that can be used to analyze transient state of the signal [1]. In [2], the internal fault has been modelling using both finite element and further analyze by DWT. The result has shown that fault in primary side does not affect secondary winding, while fault on secondary winding can affects both sides. In [3]-[4], the effect of internal, external fault, inrush current and mechanical defects in transformer on protection scheme has been discussed. In addition, classification methodology using DWT and Decision Tree (DT) and Artificial Neural Network (ANN) also been proposed. Another application of DWT on transformer protection is discussed in [5]. The proposed methodology used boundary wavelet coefficient energy that can be used in real-time protection with satisfy performance. The characteristic of wavelet coefficient in case of interturn fault using software simulation has been proposed [6]. In the recent research, only the characteristic of winding to ground fault has been analyze by using experimental setup [7]

From the many research and study that has been reviewed, it has shown that DWT based algorithm has potential to be used in fault diagnosis in transformer with high accuracy and fast response. However, the effect of fault on characteristic, and behavior of wavelet coefficient that will be used to construct those algorithms has little study to cover on. So, this paper proposed study and analysis on wavelet coefficient characteristic when internal (interturn fault) and external fault (fault on transmission line) occurring in transformer in both high and low frequency component. The study characteristic using experimental setup that consist of transmission line connected to transformer. This system has been modelled after actual system part of Thailand's transmission system. The result and methodology can be used to further improved protection scheme or fault diagnosis algorithm that based on DWT in the future.

2. Wavelet analysis

The internal and external fault signal has been simulated by using transmission line connected to transformer experimental setup that modeled after actual part of Thailand's transmission system. The wavelet coefficient study process can be illustrated as diagram in Fig. 1. Measurement on both primary and secondary winding has been done in order to obtain current during fault occurrence in both fault phase and normal phase. This signal will be used to calculate differential current. After that DWT will be applied on differential current in order to extract High and Low frequency component from the signal in both fault and normal phase. Coefficient value from high and low frequency component will be used for characteristic analysis under faulty condition.

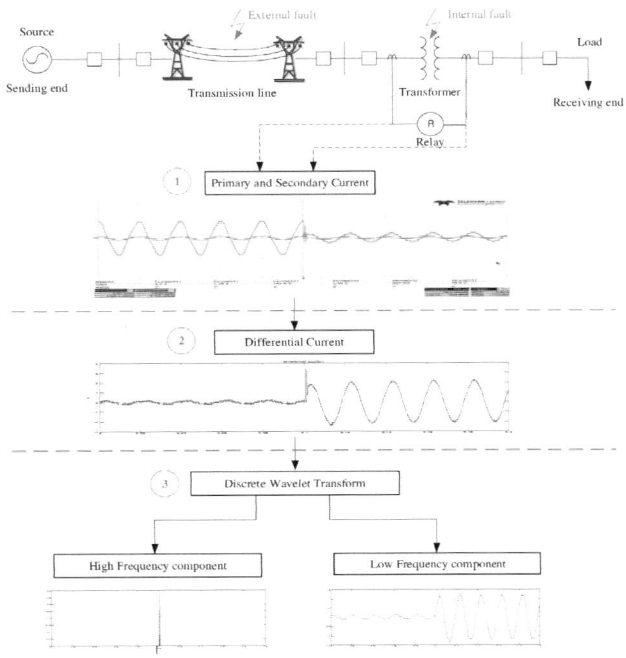

Figure 1. Overall diagram of internal and external fault analysis.

Characteristic of discrete wavelet transform under fault condition has been perform in this research by divide case study into internal and external fault case as followed.

IOP Conf. Series: Earth and Environmental Science **127** (2018) 012008 doi:10.1088/1755-1315/127/1/012008

In case of internal fault, the case studies are focus on interturn fault due to high possibility of occurrence in transformer. The research has varied width of gap between winding that interturn fault occur from 10% (44V-88V in primary winding and 22V-44V in secondary winding) to 90% (44V-440V primary winding and 220V-440V in secondary winding) with 10% interval in both primary winding and secondary winding in order to evaluate the effect of fault location on coefficient value.

In case of external fault, the case studies are focus on single line to ground fault on transmission line that connected to transformer. The location of fault on transmission line has been varied from 30% to 70% length of transmission line with 10% interval in order to take effect of location in to consideration

3. Result

In case of internal fault, the characteristic of wavelet coefficient in both normal and fault phase can be summarize as shown in Fig. 2. For internal fault on primary winding, the coefficient of low and high frequency component in Fig 2 (a) and (b) respectively, shown that coefficient in fault phase is higher than normal phase significantly in both low and high frequency component. Fault position influences coefficient with coefficient increasing as width of interturn fault increase due to rise in differential current amplitude.

For internal fault on secondary winding, the coefficient of low and high frequency component in Fig.2 (c) and (d) respectively reveal the similar trend when compare with fault on primary winding. However, wavelet coefficient in high frequency component is lower due to fault in secondary winding cause rise in current in both primary and secondary winding. This result in low amplitude for differential current.

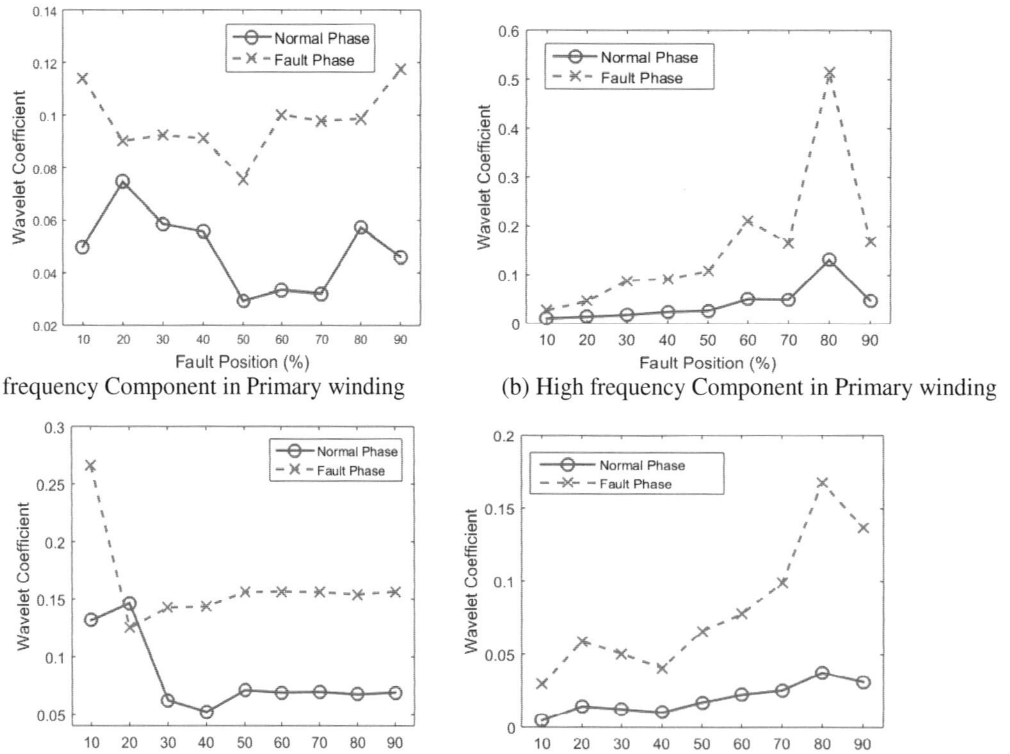

(a) Low frequency Component in Primary winding (b) High frequency Component in Primary winding

(c) Low frequency Component in Secondary winding (d) High frequency Component in Secondary winding

Figure 2. Wavelet coefficient on fault and normal phase in case of internal fault.

4. Conclusion

This paper proposed study on wavelet coefficient characteristic in both high and low frequency component when internal and external fault occurring in transformer using experimental setup on laboratory level. The result can be summarize as followed

- For internal fault that occur in transformer, the wavelet coefficient of fault phase is higher than normal phase. However, fault on primary winding has higher coefficient value due to high amplitude differential current. In addition, Fault position has significant impact on coefficient value because width of interturn fault gap has direct effect on amplitude of differential current
- For external fault that occur in transmission line that connected to transformer, the wavelet coefficient in normal and fault phase does not differentiate much with normal phase a little bit higher due to external fault affect amplitude of differential current in fault phase. Fault location on transmission line does not have significant influence on wavelet coefficient.

This characteristic of wavelet coefficient from both internal and external fault in transformer has distinguishing characteristic that can be used to construct protection scheme or fault diagnosis algorithm in the future.

5. References

[1] D.C. Robertson, O.I. Camps, J.S. Mayer, and W.B. Gish, "Wavelets and electromagnetic power system transients," in *IEEE Transactions on Power Delivery*, vol. 11, no. 2, pp. 1050-1058, Apr 1996.

[2] N. Y. Abed and O. A. Mohammed, "Modeling and Characterization of Transformers Internal Faults Using Finite Element and Discrete Wavelet Transforms," in *IEEE Transactions on Magnetics*, vol. 43, no. 4, pp. 1425-1428, April 2007.

[3] S. Bagheri, Z. Moravej and G. B. Gharehpetian, "Effect of transformer winding mechanical defects, internal and external electrical faults and inrush currents on performance of differential protection," in *IET Generation, Transmission & Distribution*, vol. 11, no. 10, pp. 2508-2520, 7 13 2017.

[4] S. bagheri; Z. Moravej; G. B. Gharehpetian, "Classification and Discrimination among Winding Mechanical Defects, Internal and External Electrical Faults and Inrush Current of Transformer," in *IEEE Transactions on Industrial Informatics*, vol.PP, no.99, pp.1-1, 28 June 2017.

[5] R. P. Medeiros, F. B. Costa and K. M. Silva, "Power Transformer Differential Protection Using the Boundary Discrete Wavelet Transform," in *IEEE Transactions on Power Delivery*, vol. 31, no. 5, pp. 2083-2095, Oct. 2016.

[6] J. Klomjit and A. Ngaopitakkul, "Behaviour of interturn fault in power transformer winding using high frequency components of discrete wavelet transform," *2012 15th International Conference on Electrical Machines and Systems (ICEMS)*, Sapporo, 2012, pp. 1-6.

[7] J. RumMdkarn and A. Ngaopitakkul, "Behavior analysis of winding to ground fault in transformer using high and low frequency components from discrete wavelet transform," *2017 International Conference on Applied System Innovation (ICASI)*, Sapporo, 2017, pp. 1102-1105

Acknowledgment

The authors wish to gratefully acknowledge financial support for this research (No. KREF156001) from King Mongkut's Institute of Technology Ladkrabang Research fund, Thailand.

IOP Conf. Series: Earth and Environmental Science **127** (2018) 012009 doi:10.1088/1755-1315/127/1/012009

GIS as a vital tool for Environmental Impact Assessment and Mitigation

Koorosh Gharehbaghi and Christina Scott-Young

RMIT University, Melbourne, Australia

koorosh.gharehbaghi@rmit.edu.au

Abstract. Environmental Impact Assessment (EIA) is a course of action which provides information to various stakeholders such as planners and relevant authorities about the planned development and its subsequent effects of the environment and the immediate ambiances. Furthermore, the EIA and mitigation are the inclusive process of collecting, analyzing information and the determination of the application for development or construction approval, which could be accessible by the concerned communities and organizations. Although the set regulations of EIA and mitigation vary from jurisdictions, they are, however, very precise and need to be integrated with the specific geographical data. In addition, the Geographical Information System (GIS) is a software intended to encapsulate and present all types of physical, biological, environmental, ecological and geological information. Conversely, GIS is the integration of statistical analysis and information technology, and can also be further broken down into two different categories of; Topological Modelling and Map overlay. To ensure that the EIA and mitigation are receptive the GIS will provide the decisive apparatus. Using GIS not only improves the overall EIA and mitigation process, but also provides valuable mapping strategies, including holistic environmental system approach. Accordingly, the main objective of this paper is to discuss the importance of the GIS and Environmental Data integration progression, to further enhance the overall EIA and Mitigation processes.

1. Introduction

Shielding and enhancing the environment is becoming an obligation rather than an excessiveness [1]. Rapid urbanization in the developing world not only has endangered the environment but also deteriorate it significantly [2]. Accordingly rigorous Environmental Impact Assessment (EIA) needs to be carefully carried out to ensure the overall ecosystem is vigilantly protected. EIA is a process and practice which bestow environmental and related information to the necessary populous. Furthermore, effective EIA needs to carefully consider some of the mitigation matters, including, Pursuing a strategic approach [2]. Moreover, the Pursuing a strategic approach includes the consideration of Impact Statement incorporating environmental assessment of; noise and vibration; groundwater and geology; soils, contamination and water quality; and sustainability factors [3].

All of these environmental strategies, including to avoid, mitigate and manage potential impacts also needs to be identified and developed as a part of the Environmental Impact Statement. According the Environment Impact Assessment also needs to consider delivering effective environmental, social and economic improvements. Consequently, sustainable ecological measures too need to be established to reduce environmental impacts such as; Waste Management, Land use integration, and Biodiversity conservation. These impacts need to be carefully integrated holistically to represent both

Environmental and Sustainability issues as a part of the Receptive Environmental Impact Assessment and Mitigation [4].

2. Receptive Environmental Impact Assessment and Mitigation

As already discussed EIA is a process and practice which deliver information to local authority planners, authorized bodies, other interested parties about intended development. EIA is furthermore, the complete process of retrieving, and analyzing information for the purpose of determination of the application for any development or construction approval [5]. The set regulations vary from state to state, however, they are very precise and cannot be disregarded [6]. The general aim of receptive EIA and mitigation is to ensure not only a cohesive impact assessment, but also ensuring that all the involved and relevant parties are successfully aware of the overall environmental implications [4]. Moreover, this receptive EIA and mitigation process not only needs to cover the main alternative environmental studies, but also provide a non-technical summary, to represent the overall environmental finding and recommendations.

In addition, the receptive EIA and mitigation needs to carefully portray and depict some of the more important Environmental and Sustainability issues, including; Ecological conditions such as condition assessment and any deterioration of the existing species [6]. To further assist with the receptive EIA, mitigation and their application the Geographical Information System (GIS) presents an excellent apparatus [7].

3. GIS

As already noted, GIS is a software intended to encapsulate, store, analyze, manage and present all types of physical, biological, environmental, ecological and geological data. Conversely, GIS is the integration of statistical data and analysis together with the Information Management utilization. For application purposes, GIS can then be broken down further into two different categories; Topological Modelling and Map overlays [8]. While the Topological Modelling involves Measurement and Mapping methodologies, Map overlays, on the other hand, shows specific overlays and restrictions implemented on the land.

Furthermore, GIS is employed in feasibility studies and economic advancements, site suitability analysis and determination. As GIS technology is advancing rapidly, it presents a manifesto for more efficient and effective planning and decision making. Accordingly, GIS not only provides mapping and visualization techniques but also illustrates the spatial analysis such as aerial representations. Moreover, GIS is being utilized ubiquitously to facilitate genuine issues and confronting additional concerns. Within EIA and Mitigation compass, GIS is exploited to sustain the emerging trend of Environmental implications by closely monitoring any detrimental environmental changes.

As [9] correctly conversed, some of the main benefits of using a GIS include easy tacking and recording the geographical changes together with effectively administrating geographic space in order to plan a course of action. For Receptive Environmental Impact Assessment and Mitigation, these benefits ensure that GIS creates a specific Environmental and physical geography mapping. This important mapping process also creates a systematic economic and development overview such as urban and transportation plotting schematics. This in all generates receptive strategies, which could be utilized as a part of the comprehensive Environmental Systems approach.

4. Environmental Systems Integration

From the above discussions, it is evident that environmental system approach is a major challenge for any EIA and mitigation process. Thus, the disposition and severity of environmental issues together with the characteristics of probable intervention approaches depends on many ecological factors. Some of these factors include scopes and magnitude of environmental problems together with the most appropriate remedies available. These remedies in-turn create the most relevant and appropriate level of responsibility and decision-making needed to solve various environmental problems [10].

While Environmental system may consist of Ecology, Environmental law, and land resources, it is the behaviour and interrelationships of such issues, which are of a great focus. Such behaviour and interrelationship may also include Air and water quality, pollution, and other global changes; all of which require a significant level of analysis, planning and decision making [11].

Accordingly, the overall environmental system approach, including the technical urban environmental management also needs to be carefully contemplated and integrated. This consists of Solid Waste Management, Water, Sanitation and Air Quality Management, together with and specific Environmental Protect Agency (EPA) requirements. In addition, the creation of an integrated urban environmental management plan that exemplifies short and long-term goals and phased targets for meeting these goals. All of these matters need to be carefully considered and implemented within the effective Environmental System Integration, which is illustrated in figure 1.

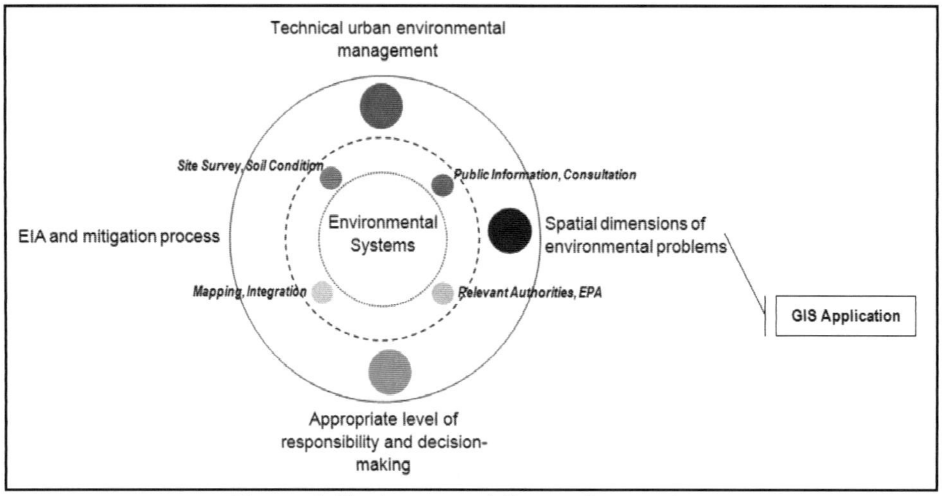

Figure 1. Environmental System Integration.

As it can be noticed, an effective Environmental System involves the integration of all the main considerations, including EIA and the mitigation process. In addition, each of these main considerations is further expanded into sub-levels such as mapping process. Moreover, a key component of this amalgamation is the GIS and Environmental data integration.

4.1 GIS and Environmental Data Integration

The relationship of GIS and Environmental Data Integration are a unique association which must preserve a premeditated roadmap for future development of any environment. This association also needs to carefully consider the involvement of innovative technologies and the growing difficulties of effective and successful environmental management processes. In addition, this connection would further offer a configurable phase to assemble, analyze, and present environmental information [12].
The integration of GIS and Environmental Data, needs to operate proactively by surveying, up-loading, configuring and validating data. Appropriately, this integration would then allow for environmental data to be accessed anywhere, anytime and be able to be updated. In addition, the GIS and Environmental Data integration also need to visualize data, explore and report information, model outputs, and other records that change the response workflow.

More importantly, the relationship of GIS and Environmental Data integration will provide for real-time analysis, and the complex permutation of spatial and temporal data sets. This unique GIS and Environmental Data Integration are exemplified in figure 2.

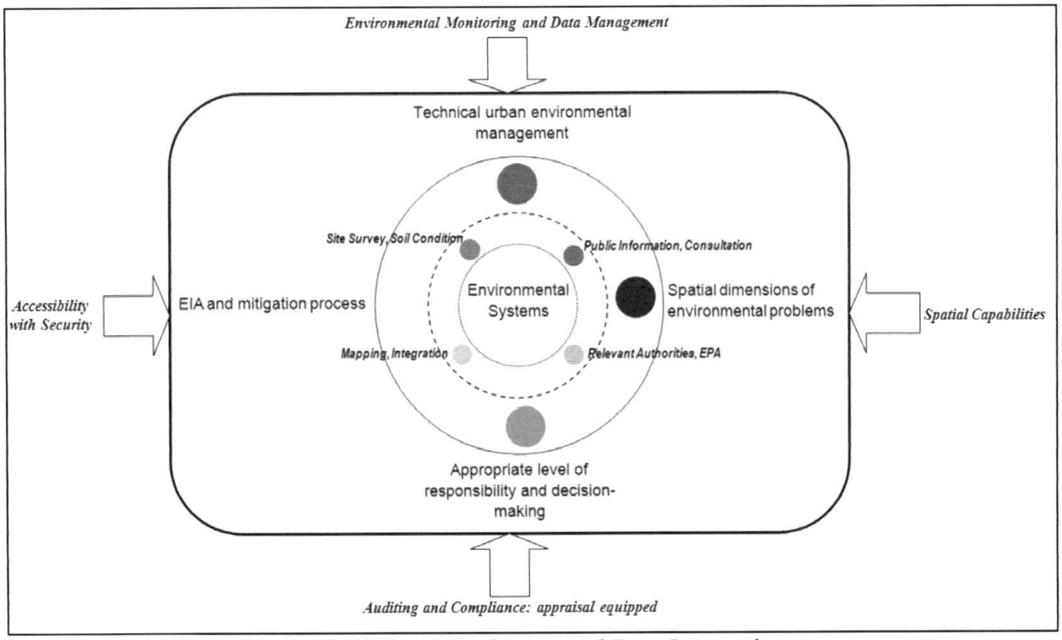

Figure 2. GIS and Environmental Data Integration.

As it can be noticed, the receptive method of GIS and Environmental Data Integration can be categorized based on:-

- Spatial Capabilities: including map's integration, and data amalgamation enabling spatial representation.
- Environmental Monitoring and Data Management: monitor and regulate information to reduce the ecological risk, and better application of monitoring programs.
- Accessibility with Security: access, validated and protect historical data and allow approved individuals to monitoring programs and admittance the stored data sites.
- Auditing and Compliance: permitting the environmental data, including historical for the auditing purposes as required by the specific jurisdictions.

A key aspect of the GIS and Environmental Data Integration is the ability to further navigate, modify, calculate data and store specific technical information, including precise schematics and photographs via customized system dashboards. This customized system dashboards will not only entail natural environment information such as Hydrography and soil geology but also Flora and Fauna facts, thus creating a responsive approach.

Nevertheless, GIS will further assist with improving the EIA through careful and optimised amalgamation of ecosystem specific information such as any short and long-term deterioration evaluation and assessment. Accordingly, such evaluation requires effective environmental strategy as a part of the precise data integration.

4.2 Effective Environmental Strategy as a part of the precise data integration

To further promote the GIS and Environmental Data Integration, valuable environmental strategies need to be sympathetically implemented. These environmental strategies include [13]:-

- Institutional urban environmental management considerations. This includes in cooperating sustainability with environmental conditions during increasing local procedures. As a part of this global sustainability attitude they also needs to be regularly updated.
- Informational urban environmental management deliberation. This comprises of contemporary Software and Information Systems (for example, spatial capabilities) as a part of the flexible and functional approach.

- Technical urban environmental management contemplation. This consists of Solid Waste Management, Water, Sanitation and Air Quality Management, together with Traffic Management. All of these are based on the relevant acts a regulation, including the AS/NZS 1547:2000 and the Commonwealth Radioactive Waste Management Act 2005.

As it can be noticed these environmental strategies require careful planning and development in alignment with the Environmental Dynamics and Sustainability factors. Accordingly, the overall planning strategies as a part of the GIS and Environmental Data integration are [14], [15] and [16]:-

- Protecting and improving the environment in both urban and rural areas through institutional strengthening. Utilizing the institutional strengthening to minimize the urban and rural impact on natural resources at the local, state and federal jurisdictions [17] and [18].
- The creation of an integrated urban environmental management that symbolizes short and long-term goals and policy reforms of the EIA and mitigation processes.
- Integrated management of environment and natural resources through informed consultation with all the relevant entities and establishing set of achievable priorities as the basis of effective EIA and mitigation.
- Integration of the nature and severity of environmental problems, to initiate a set of agreed programs to further improve the effectiveness of the EIA and mitigation processes.

In addition, as the basis of such date integration, Sydney Metro project in Australia can be exploited. For this mega transportation project, the GIS was a vital tool for assisting with the EIA data integration through:-

- Determining the unique natural features of Sydney and its urban areas. This included the determination of the local ecosystem characteristics.
- Establishing the diverse spatial dimensions of Sydney environmental problems and the subsequent severity of various impact.
- Deciding the actual roles of local actors and authorities involved in this project.

5. Conclusion

Environmental Impact Assessment and mitigation are the complete process of collecting, analyzing information and the determination of the application for development or construction approval, which could be accessible by the concerned communities and organizations. Moreover, GIS is a software intended to encapsulate, store, analyze, manage and present all types of physical, biological, environmental, ecological and geological data.

To ensure that environmental impact assessment and mitigation are receptive the GIS will present the crucial tool. Using GIS not only improves the overall EIA and mitigation process, but also provides valuable mapping strategies, including holistic environmental system approach. Accordingly, this paper carefully discussed the main advantages of GIS to further present responsive EIA and Mitigation processes. In doing so, this paper also presented and discussed the receptive elements of the Environmental System Integration. Finally, this paper also conversed the significance of the GIS and Environmental Data Integration.

6. References

[1] Vezzoli C, and Manzini E, "Design for environmental sustainability", Springer, 2010.

[2] Faiza, A, Suleman L, Shah S and Farhana, S, "Incorporating permaculture and strategic management for sustainable ecological resource management", *Journal of Environmental Management*, 1 September 2016, Vol.179, pp.31-37

[3] Andres B, Manuele M; Pierre-Olivier R; Cecile B; and Zwichy M, "A proposal to measure absolute environmental sustainability in life cycle assessment", *Ecological Indicators*, 2016, Vol.63, p.1(13).

[4] Gharehbaghi K. and Raso V, "Architecture Anthology I: Sustainable Design - Optimisation of Infrastructure Systems for Melbourne", pages 73 - 84, ATINER, 2015.

[5] Gharehbaghi, K, "On-Site Engineering Information Systems (EIS) for Building and

Construction Projects", International Journal of Civil Engineering and Architecture Vol 4, Issue 1, pp. 8 - 13, 2016.

[6] Cao Giacomo and Orrù Roberto, "Current environmental Issues and Challenges, SpringerLink, 2014.

[7] Casparian A and Sirokman G, "Chemistry for environmental engineering," Momentum Press, 2016.

[8] Chmielowski Wojciech, "Fuzzy Control in environmental engineering", Springer eBooks, 2016.

[9] Gharehbaghi K, "Infrastructure Asset Optimisation in Local Governments: Australia Study", International Journal of Civil, Structural, Environmental and Infrastructure Engineering Research Development, Volume: 4, Issue: 6, Page: 33 – 42, Dec 2014.

[10] Byrne J, Sipe N and Dodson J, "Australian environmental planning: challenges and future prospects", Routledge, 2004.

[11] Young R, "The Oregon Way: Planning a Sustainable Economy in the American West", Journal of Planning Education and Research, Spring 2016, Vol.36(1), p.76

[12] Gharehbaghi K and Sagoo, A, "Environmental Implications of the Sydney Metro: A Case Study", International Proceedings of Chemical, Biological and Environmental Engineering 97(3), 2016, pp. 15-21.

[13] Hardisty P, "Environmental and economic sustainabilit", CRC Press/Taylor and Francis, 2010.

[14] Lawrence D, "Environmental Impact Assessment: Practical Solutions to Recurrent Problems", Hoboken : Wiley-Interscience, 2003.

[15] Tyler G and Spoolman S, "Sustaining the earth", Cengage Learning", Eleventh edition. 2015.
UNE EN ISO 14040:2006 - Environmental Management - Life Cycle Assessment - Principles and Framework.

[16] Gharehbaghi, K and Georgy, M, "Utilization of Infrastructure Gateway System (IGS) as a Transportation Infrastructure Optimization Tool", International Journal of Traffic and Transportation Engineering, SAP, USA, 4(1), pp. 8-15, 2015.

[17] AS/NZS ISO 14042:2001 - Environmental management - Life cycle assessment - Life cycle impact assessment (Environmental Impact Assessment).

[18] EPA (USA), "Sustainability concepts in decision-making: tools and approaches for the US Environmental Protection Agency," National Academies Press, 2014.

IOP Conf. Series: Earth and Environmental Science **127** (2018) 012010 doi:10.1088/1755-1315/127/1/012010

Waste diminution in Construction projects: Environmental Predicaments

Koorosh Gharehbaghi and Christina Scott-Young

RMIT University, Melbourne, Australia

koorosh.gharehbaghi@rmit.edu.au

Abstract. Waste diminution in construction projects is not only a behavioural issue, but also an energy consumption and reduction concern. With construction waste equating to the significant amount of exhausted energy together with increased pollution, this contributes to a series of environmental predicaments. The overall goal of construction solid Waste Management is to collect, treat and dispose of solid wastes generated by project activities in an environmentally and socially satisfactory manner, using the most economical means available. As cities expand, their construction activities and consumption patterns further drive up the solid waste quantities. Governments are usually authorized to have responsibility for providing solid Waste Management services, and various administrative laws give them exclusive ownership over the waste produced. In addition, construction waste processing can be further controlled and minimized according to specialized authorities such as Environmental Protection Agencies (EPA) and their relevant acts and regulations. Moreover, a Construction Environmental Management Plan (CEMP) can further control the treatment of waste and therefore, reduce the amount produced. Key elements of a CEMP not only include complying with relevant legislation, standards and guidance from the EPA; however, also to ensuring that there are systems in place to resolve any potential problems associated with site activities. Accordingly, as a part of energy consumption and lessening strategies, this paper will discuss various effective waste reduction methods for construction projects. Finally, this paper will also examine tactics to further improve energy efficiency through innovative construction Waste Management strategies (including desirability rating of most favourable options) to promote the lessening of overall CO^2 production.

1. Introduction

Waste diminution in construction projects is not only a behavioral concern, but is also an energy consumption and lessening apprehension. The overall goal of waste diminution in construction projects is not only to collect, but also to effectively treat and dispose of wastes in an environmentally and socially satisfactory manner, using the most economical means available. As more construction projects are being built, greater quantities of waste are being produced. Moreover, increased traffic congestion adversely affects the productivity of the solid Waste Management. This productivity loss is aggravated by longer hauls since increasingly open land for disposal is further and further away from urban centers [1]. Accordingly, this will also increase the overall cost of e waste diminution. In developed countries, it is common to spend 5-7 percent of their available budget on waste diminution [2]. This high proportion of the cost is due to a comprehensive and thorough waste diminution strategies and recycling practices [3].

The sustainability of waste diminution and subsequent practices is a valuable solution in providing

IOP Conf. Series: Earth and Environmental Science **127** (2018) 012010 doi:10.1088/1755-1315/127/1/012010

an effective Waste Management service [4]. Such Waste Management services also satisfy the needs of the end users [3]. Accordingly, a key factor of sustainable solid waste diminution is appropriate strategic planning, as well as adhering to associated acts and regulations. Such acts and regulations are also dynamics of useful analytical tools [5]. On the other hand, various pollution reduction strategies also further enhance the sustainability of waste diminution [3] and [6]. These pollution reduction strategies support global needs and provide appropriate calculations on how to achieve sustainable waste diminution and management procedures. Waste Management procedures thus are crucial components of waste diminution in construction projects.

2. Waste diminution in Construction projects

The crucial components of waste diminution in construction projects not only include effective and appropriate Waste Management procedures; but also a sustainable framework [7]. According to [8] and [9], sustainable waste diminution in construction projects includes):-

1. Strategic planning and preparation. Undertaking state and federal strategic planning to design and implement integrated waste systems, which are responsive to dynamic demographic growth.
2. Implementation of a comprehensive framework. An inclusive policy outline is produced at the national, state, and local government level to connect, environmental, social, economic and other community-based policies.
3. Institutional capacity building. The institutions that provide the service typically need to be restructured so that they are more accountable and transparent to the stakeholders which they serve. The creation of specific departments managed by qualified engineers who are trained in systems design and operation's rationalization.
4. Economic impact. Implement specialized waste transfer stations and long-distance haulage, to reduce the overall cost of collecting, transferring and disposing of construction waste. In addition, utilize government cost-sharing arrangements such as the creation of fund subsidies to further motivate new waste reduction technologies.
5. Analysis of technology choices. To determine the most cost-effective Waste Management systems, carefully analyze various technology-based alternatives. Such technology-based alternatives include, improved computer-aided machines, which are not only faster in processing construction waste, but also are capable of minimizing waste through recycling processes. In evaluating such machines, it is conventional to conduct time-and-motion studies of existing systems and tests of various operating methods.

To further improve waste diminution in construction projects, managers need to first differentiate waste produced from low-rise compared to high-rise buildings. Table 1, represents the material waste differentiation in the USA [10], UK [11] and Australia [12].

Table 1. A cross-national comparison of waste material production in high and low rise construction.

Countries	Waste Material	High Rise Buildings	Low Rise Buildings
USA	Metal	18%	8%
	Concrete	28%	11%
	Timber	8%	18%
	Plastic	7%	8%
	Ceramic	8%	6%
	Others	*31%*	*49%*
UK	Metal	13%	8%
	Concrete	20%	10%
	Timber	8%	11%
	Plastic	15%	15%
	Ceramic	8%	6%
	Others	*36%*	*50%*
Australia	Metal	17%	10%
	Concrete	24%	13%

	Timber	8%	18%
	Plastic	8%	8%
	Ceramic	8%	6%
	Others	*35%*	*45%*

While [10] argued that the main waste material in the USA construction industry for high-rise is concrete and subsequent low-rise is timber, [11] claimed that for the UK, concrete is the main waste material for both high and low rise buildings. Conversely, [12] also maintained that for the Australian construction industry concrete, and timber are the main waste materials.

Due to its composition, timber can easily be reprocessed into reusable materials. Recycling. concrete, on the other hand, is a much more complex and costly process. Figures 1a and 1b represent the collective (USA, UK and Australia) average of waste materials for both low and high rise buildings, derived from Table 1.

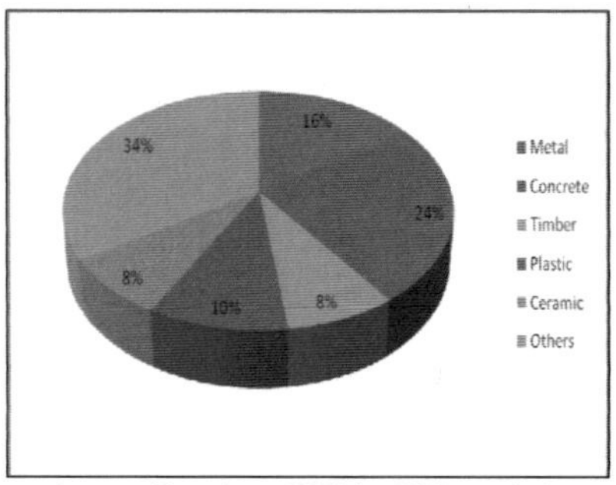

Figure 1a. collective (USA, UK and Australia) waste materials of high rise buildings.

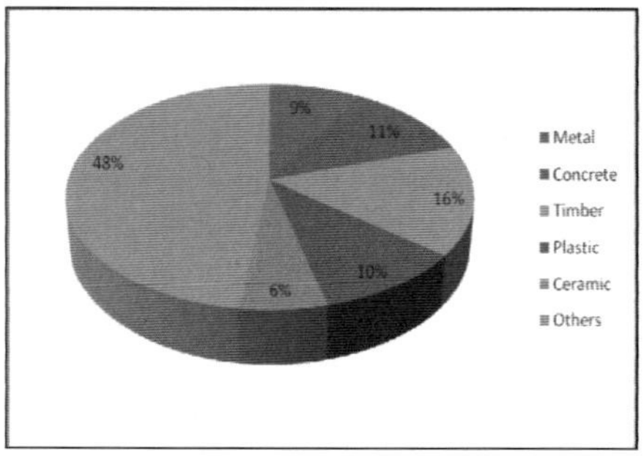

Figure 1b. collective (USA, UK and Australia) waste materials of low rise buildings.

Compared to other materials, both concrete and metal account for a high percentile of waste for both low and high rise buildings. Although low rise buildings have higher timber waste, this material is easily recyclable and is also eco-friendly. It is thus concrete and metal that could cause significant environmental consequences due to the greater difficulty and energy usage in recycling them. As [13] correctly argued, one of the most important issues during the Waste Management process is the overall environmental predicaments. Since construction projects produce a high amount of waste, the environmental predicament is a fundamental issue to contemplate, when considering waste diminution in construction projects.

3. Environmental Predicaments

Matching environmental and economic endeavors requires a Waste Management strategy that facilitates ecosystems, protects land and enhances cultural resources [14]. Consequently, waste diminution in construction projects generates environmental dilemmas and predicaments. Figure2 below represents a generic construction Waste Management outline, with environmental predicaments highlighted.

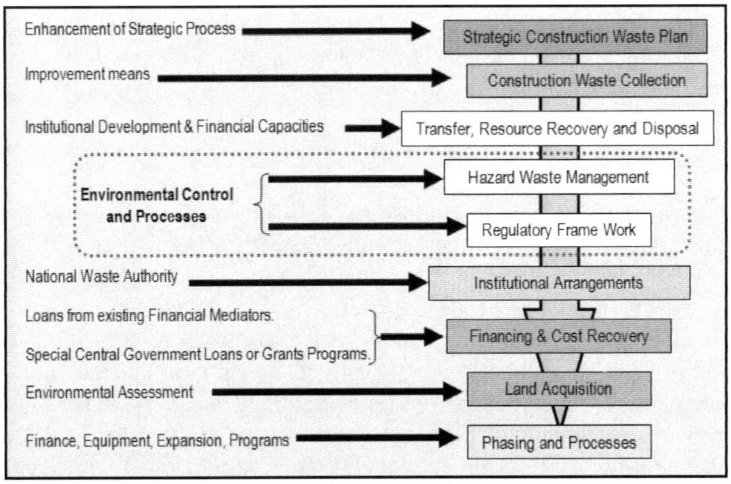

Figure 2. Generic construction Waste Management outline.

As can be noticed, as a part of the environmental predicaments, the actual control and processes need to be carefully integrated into the Waste Management and regulatory outline phases. Since such environmental predicaments are integral components of any effective waste diminution plans for construction projects, they need to be cautiously premeditated and implemented. Drawing on the work of [15], [16] and [17] such environmental predicaments include:-

- How to effectively reduce the amount of waste produced. This includes the reduction of unnecessary waste through the diminution of redundant construction consumptions.
- How to efficiently reuse the waste without reprocessing or remanufacturing. This comprises the reduction of construction waste through reusing packaging, and other useable materials.
- How to successfully recycle the waste without significant pollution production.
- How to proficiently recover the energy for the purpose of heat, electricity and fuel.
- How to resourcefully dispose of the waste through appropriate landfill processes and sites for the unrecyclable waste.

The overall environmental predicaments of waste diminution in construction projects is provided in Figure 3 below.

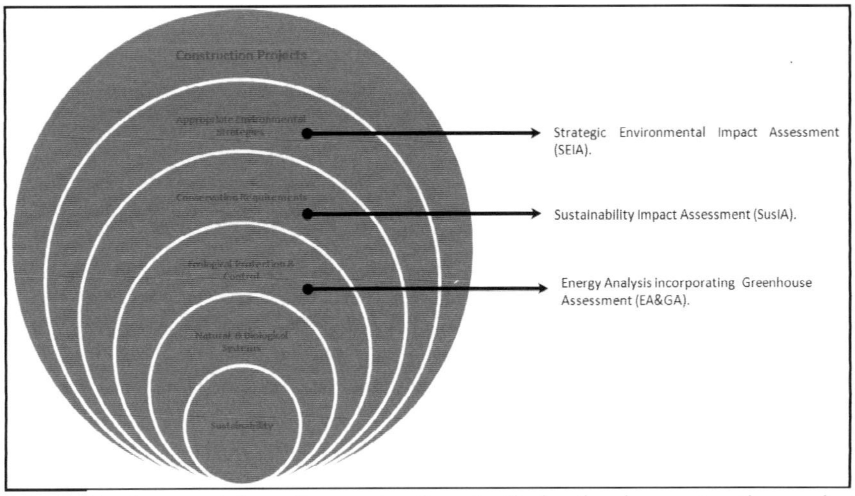

Figure 3. Environmental predicaments of waste diminution in construction projects.

As can be noticed, the environmental predicaments of waste diminution in construction projects consist of three main factors:-

1. Energy Analysis incorporating Greenhouse Assessment (EA&GA). While Energy Analysis provides findings using fields such as Portfolio Impacts Analysis; Greenhouse Assessment, on the other hand, presents methodologies for calculating CO2 emissions [18]. Accordingly, the combination of Energy Analysis and Greenhouse Assessment should further improve the evaluation of alternatives, in particular, for construction material manufacturing methods.
2. Strategic Environmental Impact Assessment (SEIA). This is a process to support decisions in the context of sustainable environment strategies. It further assists with waste policy-making and promoting such guidelines via positioning key environmental milestones together with integrating the green economy [19].
3. Sustainability Impact Assessment (SusIA). This can further assist with governance procedures through improved transparency. It can also enhance decision making because of systematically assessing economic, environmental and social impacts [20].

To further examine these three main factors, their occurrence in US, UK and Australia construction industries were also investigated. Table 2 provides the comparison of the local, state and federal environmental jurisdictions in these countries.

Table 2. Cross-national comparisons of local, state and federal environmental jurisdictions.

Countries	EA&GA	SEIA	SusIA
USA	Local ○ State ○ Federal ●	Local ● State ○ Federal ○	Local ○ State ● Federal ●
UK	Local ○ State ○ Federal ●	Local ○ State ● Federal ○	Local ○ State ○ Federal ●
Australia	Local ○ State ○ Federal ●	Local ● State ● Federal ○	Local ○ State ● Federal ●

As it can be observed, while the EA&GA are applied at the federal level in each country, the SEIA relates to either the local or state level, (or both as in Australia). The SusIA on the other hand, pertains at either the federal jurisdiction in the UK, or at both federal and state levels in the US and Australia. Such variances in the mandates could thus create governing dilemmas. Once again, such governing complexity could also cause additional environmental predicaments especially for construction projects. It is thus fundamental to ensure that all the three main environmental factors (EA&GA, SEIA and SusIA), are treated holistically rather than individually. Even though, these three main environmental factors have different systems and ecological foci, they need to be carefully integrated within the Construction Environmental Management Plan (CEMP). In addition, the three main factors of waste diminution in construction projects need to be carefully applied in all the applicable jurisdictions [21] and [22].

4. Conclusion

The overall goal of construction solid waste diminution is to carefully manage the excess in an environmentally and socially suitable manner. Although an appropriate CEMP would further control the treatment of the waste, it is ultimately the three main factors of waste diminution in construction, which further refines such processes. As a part of the waste diminution for the construction projects, this paper first examined the most common waste materials in the USA, UK and Australia construction industries. Moreover, this paper determined that environmental predicaments of waste diminution in construction consist of three main factors (EA&GA, SEIA, and SusIA). Accordingly, all of these three factors were investigated to prompt recommendations for improved environmental processes for waste diminution in construction projects.

References

[1] Gharehbaghi, K. and Raso, V., (2015), "Architecture Anthology I: Sustainable Design - Optimisation of Infrastructure Systems for Melbourne", pages 73 - 84, ATINER publications.

[2] Woosik, J.; Hyun-Woo, Y. and Seung Heon, H., (2015), "Quantitative Decision Making Model for Carbon Reduction in Road construction Projects Using Green Technologies", Journal of Sustainability, Vol.7(8), pp.11240-11259.

IOP Conf. Series: Earth and Environmental Science **127** (2018) 012010 doi:10.1088/1755-1315/127/1/012010

[3] Young, R., (2016), "The Oregon Way: Planning a Sustainable Economy in the American West", Journal of Planning Education and Research, Vol. 36(1), pp.60-76.
[4] Gharehbaghi, K, (2017), "Modular High-rise Construction as an alternative Building System", International Journal of the Constructed Environment, Vol 8, Issue 3, pp 15-25.
[5] Gharehbaghi, K, (2015),"Advancements in Concrete Technology in Australia: Geo-Polymer Concrete", International Journal of the Constructed Environment, Vol 7, issue 1, pp 19-29.
[6] Acton, A., (2013),"Issues in Global Environment: Pollution and Waste Management", ScholarlyEdition, Atlanta, USA.
[7] Li, R., (2015), "Construction Safety and Waste Management: An Economic Analysis: An economic analysis", Springer, New York.
[8] Elliott, M., (2014), "Environmental Impact Assessment in Australia - Theory and practice", 6th edition, The Federation Press, NSW, Australia,
[9] Vezzoli, C. and Manzini, E., (2010), "Design for environmental sustainability", Springer.
[10] Chen, Z. and Li, H. (2006), "Environmental Management in Construction: A Quantitative approach", Taylor and Francis, USA.
[11] Oyedele, L.; Regan, M.; Ahmed, A. and Elnokaly, A., (2013), "Reducing waste to landfill in the UK: identifying impediments and critical solutions", World Journal of Science, Technology and Sustainable Development, Vol.10(2), pp.131-142.
[12] Udawatta, N., Zuo, J., Chiveralls, K., and Zillante, G., (2015), "Improving Waste Management in Construction Projects: An Australian Study", Journal of Resources, Conservation and Recycling, vol 101, pp. 73-83.
[13] Australian Department of Environment and Energy, www.environment.gov.au/
[14] Bernstein, J., (1993), "Land use considerations in urban environmental management", Transportation, Water and Urban Development department, The World Bank, Urban No. UE-10.
[15] Byrne, J.; Sipe, N.; and Dodson, J., (2004), "Australian environmental planning : challenges and future prospects", Routledge, London.
[16] Cao, G. and Orrù, R., (2014), "Current environmental Issues and Challenges", Springer, USA.
[17] Chmielowski, W., (2016), "Fuzzy Control in environmental engineering", Springer Books.
[18] Faiza, A.; Suleman, L.; Shah, S. and Farhana, S., (2016), "Incorporating permaculture and strategic management for sustainable ecological resource management", Journal of Environmental Management, September, Vol.179, pp.31-37.
[19] Gangolells, M.; Casals, M. ; Forcada, N. and Macarulla, M., (2014), "Analysis of the implementation of effective Waste Management practices in construction projects and sites", Journal of Resources, Conservation & Recycling, Vol.93, pp.99-111.
[20] Valli Manickam, A., (2011), "Environmental Impact Assessment Methodologies", 2nd edition, CRC Press.
[21] Williams, P., (2013), "Waste Treatment and Disposal", 2nd edition, Wiley, USA.
[22] EPA (USA), "Sustainability concepts in decision-making : tools and approaches for the US Environmental Protection Agency", National Academies Press, USA,

IOP Conf. Series: Earth and Environmental Science **127** (2018) 012011 doi:10.1088/1755-1315/127/1/012011

Credit Risk Evaluation of Large Power Consumers Considering Power Market Transaction

Li Fulin[2], Xu Erfeng[3], Sun ke[1], Liu Dunnan[3] and Shen Shuyi[4]

[1] State Grid Zhejiang Electricity Power Company, 8 Huanglong Road, Hangzhou, China
[2] Zhejiang Huayun Electric Power Engineering Design Consulting Company, 1 Huadian Road, Hangzhou, China
[3] North China Electric Power University, 2 Beinong Road, Beijing, China
[4] State Grid Zhejiang Economic Research Institute, 1 Nanfu Road, Hangzhou, China

Corresponding Author's E-Mail Address: 18810905481@163.com

Abstract. Large power users will participate in power market in various forms after power system reform. Meanwhile, great importance has always attached to the construction of the credit system in power industry. Due to the difference between the awareness of performance and the ability to perform, credit risk of power customer will emerge accordingly. Therefore, it is critical to evaluate credit risk of large power customers in the new situation of power market. Firstly, this paper constructs index system of credit risk of large power customers, and establishes evaluation model of interval number and AHP-entropy weight method.

1. Introduction

Power industry has always attached great importance to construction of credit system. With the development of power system reform, the types of power market transactions will become more diversified while power market environment will more variable and complex. Therefore, institution of power market regulation should evaluate credit risk of all market subjects urgently, which will provide decision-making basis for effective prevention of market risk and technical support for protection of market standardized operation and healthy and orderly development.

There are many researches on credit index system of power industry. Literature [1] has proposed a customer credit evaluation index system, including the status of industry, the financial situation of enterprises, payment status and so on. Literatures [2-4] have calculated the correlation coefficient between different credit categories. Literature [5] has evaluated the credit risk of power customers based on AHP. In literatures [6-7], power customer risk evaluation model has been constructed by entropy weight method and improved entropy weight method.

In this paper, firstly index system of credit risk evaluation is constructed. Then, based on interval numbers and comprehensive weights, a credit evaluation model is set up, which provides a theoretical reference for the evaluation of large power customer credit risk.

2. Construction of Credit Risk Index System for Large Power Customers

In this paper, credit analysis of large power customers is divided into four parts: external environment, production and operation, financial analysis and transaction credit analysis. As **Table 1** shown, credit evaluation index system for large power customers consists of 4 criterion layers, 9 primary factor layers, and 23 secondary factor layers.

Table 1. Credit Risk Evaluation Index System for Large Power Customers.

Target Layer	Criterion Layer	Primary Factor Layer	Secondary Factor Layer
Credit Risk of Large Power Customers	Credit Risk of External Environment	Policies and Regulations	Government Support
			Industrial Macro-Control Policies
		Industry Operation Status	Fluctuation of Supply and Demand
			Consolidated Price Increase
	Credit Risk of Production and Operation	Operation Risk	Market Share
		Production Risk	Corporate Social Impact
			Product Qualification Rate
			Safety Accident Rate
	Credit Risk of Financial	Profitability	Return on Assets
			Net Profit Margin on Sales
			Cost Profit Margin
			Asset Liability Ratio
		Solvency	Current Ratio
		Development Ability	Main Business Revenue Growth Rate
			Net Profit Growth Rate
	Credit Risk of Transaction	Direct Purchase of Large Power Consumers	Full Payment Rate
			Time Contribution Ratio
			Contracting Rate of Electricity Sales Contract
			Actual Power Completion Rate
			Contract Performance Rate
		Demand Side Response	Interruptible Load Contract Performance Rate
			Load Control Coordination
			Electricity Check Fit

2.1. External Environment Risk
Enterprises are affected by national environment at a macro level. Besides, the operation of industry and the state related to this industry should also be considered. Therefore, this paper will analysis the project from the aspects of policies and the running state of the industry.

2.2. Production and Operation Risk
The overall level of enterprises operated well is in a leading position, and the credit risk can reduce. While the enterprises operated poorly may face the risk of losses or failures. This paper will discuss from two perspectives of the production level and operating conditions of large power customers.

2.3. Financial Credit Risk
The financial situation of large power customers reflects the enterprise management level, will also affect the future development and the competitiveness of the enterprise. The financial risk can be divided into solvency risk, profitability risk and development capacity risk.

IOP Conf. Series: Earth and Environmental Science **127** (2018) 012011 doi:10.1088/1755-1315/127/1/012011

2.4. Transaction Credit Risk

The transaction credit risk can be divided into direct trade credit risk and demand side response to the credit risk according to the type of market transactions that large power customers participate in.

3. Credit Risk Evaluation Method for Large Power Customers

3.1. Determine Index Score by Interval Number Method

3.1.1. Dimensionless Treatment for Interval Numbers. Dimensionless treatment is to eliminate the influence of dimensional difference among different indicators on evaluation.

$$Y_{Max,ij}^{P} = X_{ij}^{P} / \sum_{i=1}^{m} X_{ij}^{P}, Y_{Max,ij}^{Q} = X_{ij}^{Q} / \sum_{i=1}^{m} X_{ij}^{Q}, Y_{Min,ij}^{P} = \frac{1/X_{ij}^{P}}{\sum_{i=1}^{m} 1/X_{ij}^{P}}, Y_{Min,ij}^{Q} = \frac{1/X_{ij}^{Q}}{\sum_{i=1}^{m} 1/X_{ij}^{Q}} \tag{1}$$

3.1.2. Value Transformation for Interval Numbers. Supposing that interval numbers are subordinated to normal distribution and exchanged into evaluation value by standardization.

$$u_{ij} = \frac{X_{ij}^{P} + X_{ij}^{Q}}{2}, \sigma_{ij} = \frac{X_{ij}^{Q} - X_{ij}^{P}}{6}, \sigma_{ij} = u_{ij} + \lambda \times \left(1 - \sigma_{ij}\right) \tag{2}$$

3.2. Credit Evaluation Process Based on AHP and Entropy Weight Method (EWM)

3.2.1. Determine Objective Weight by Entropy Weight Method (EWM).

(1) Calculate the proximity between each large power customer index and the optimal value.

$$x_{ij} = x_{ij} / x_{j}^{*}, x_{j}^{*} = Max\left(x_{ij}\right), x_{ij} = x_{j}^{*} / x_{ij}, x_{j}^{*} = Max\left(x_{ij}\right), d_{ij} = \frac{x_{ij}}{\sum_{j=1}^{m}\sum_{i=1}^{n} x_{ij}} \tag{3}$$

(2) Calculate importance entropy of index.

$$E_{ij} = -d_{ij} \times \ln\left(d_{ij}\right), e_{ij} = -\frac{d_{ij}}{\sum_{j=1}^{m} d_{ij}} \times \ln\left(d_{ij} / \sum_{j=1}^{m} d_{ij}\right), e_{j} = \frac{1}{\ln(m)} \times \sum_{j=1}^{m} e_{ij} \tag{4}$$

(3) Calculate weight of index.

$$u_{j} = \left(1 - e_{j}\right) / \left(n - \sum_{j=1}^{n} e_{j}\right) \tag{5}$$

3.2.2. Determine Objective Weight by AHP.

(1) According to the experience of experts and credit rating workers' opinion, criterion indexes are ranked by importance in pairs, which form a judgment matrix.

(2) Calculate weight feature vector.

$$M_{i} = \prod_{j=1}^{n} x_{ij}, \overline{W}_{i} = \sqrt[n]{m_{i}}, W_{i} = \overline{W}_{i} / \sum_{j=1}^{n} \overline{W}_{i} \tag{6}$$

3.2.3. Calculate Comprehensive Evaluation Value.

Calculate maximum eigenvalue of matrix and get comprehensive evaluation value.

IOP Conf. Series: Earth and Environmental Science **127** (2018) 012011 doi:10.1088/1755-1315/127/1/012011

$$\varphi_{Int} = 0.5 \times \varpi_{AHP} + 0.5 \times \varpi_{EWM}, \chi = D \times \varphi_{Int} \quad (7)$$

4. Credit Rating of Large Power Customers

According to the evaluation process above, subjective and objective weight are calculated to get integrated weight of large power customers, as shown in **Table 2**.

Table 2. Interval Number of Large Power Customer Evaluation Indexes.

Target Layer	Weight	Primary Factor Layer	Weight	AHP	EWM	Integrated Weight
Credit Risk of External Environment	0.0776	Government Support	0.2275	0.0043	0.0000	0.0021
		Industrial Macro-Control Policies	0.4834	0.0091	0.0000	0.0046
		Fluctuation of Supply and Demand	1.0798	0.0204	0.0425	0.0314
		Consolidated Price Increase	2.3280	0.0437	0.0425	0.0431
Credit Risk of Production and Operation	0.2010	Market Share	0.4396	0.0211	0.0450	0.0331
		Corporate Social Impact	0.9356	0.0435	0.0451	0.0443
		Product Qualification Rate	1.4270	0.0681	0.0422	0.0552
		Safety Accident Rate	1.4270	0.0681	0.0425	0.0553
Credit Risk of Financial	0.2010	Return on Assets	0.4956	0.0142	0.0487	0.0315
		Net Profit Margin on Sales	0.8806	0.0251	0.0379	0.0315
		Cost Profit Margin	1.5491	0.0444	0.0568	0.0506
		Asset Liability Ratio	0.4956	0.0142	0.0476	0.0309
		Current Ratio	0.4596	0.0142	0.0386	0.264
		Main Business Revenue Growth Rate	1.5491	0.0444	0.0841	0.0643
		Net Profit Growth Rate	1.5491	0.0444	0.0841	0.0643
Credit Risk of Transaction	0.5205	Full Payment Rate	0.9776	0.0635	0.0425	0.0530
		Time Contribution Ratio	0.5253	0.0341	0.0446	0.0394
		Contracting Rate of Electricity Sales Contract	0.9776	0.0635	0.0429	0.0532
		Actual Power Completion Rate	1.7587	0.1139	0.0425	0.0782
		Contract Performance Rate	1.7587	0.1139	0.0425	0.0782
		Interruptible Load Contract Performance Rate	0.9776	0.0635	0.0417	0.0526
		Load Control Coordination	0.5253	0.0341	0.0448	0.0394
		Electricity Check Fit	0.5253	0.0341	0.0412	0.0376

As **Figure 1** shown, comprehensive evaluation value of large power customer 3 is the highest, followed by large power customer 2. Credit rating of these two corporates is at a high level, which indicates that they have a certain reputation in the industry with good business development, high debt credit, strong resistance to external environmental impact and low default risk.

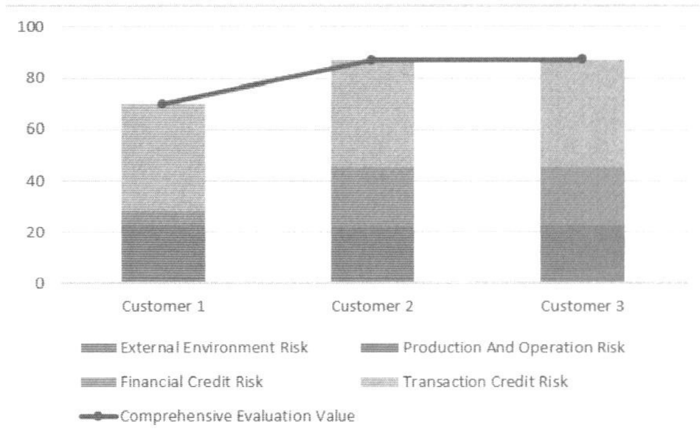

Figure 1. Evaluation Value of Large Power Customer Credit.

5. Conclusion

With the development of power system reform, credit problem in power industry will be more complicated. In this paper, a set of credit risk evaluation index and evaluation model based on comprehensive evaluation are constructed. As the results of the numerical examples shown, credit risk of large power customers is closely related to external environment risk, production and operation risk, financial credit risk and transaction credit risk. Also, we can compare and analyze the credit rating of different enterprises and the indicators to be improved. The model established in this paper provides a quantitative reference for the credit prevention of power market transactions.

6. References

[1] Hao Yang. Research on Risk Management of Electricity Power Supply Enterprises in China. Beijing, North China Electric Power University, 2014.

[2] Chen Yanchao. Study of the Credit Evaluation of Key Customer and the Risk Management of the Power Company. Beijing, North China Electric Power University, 2015.

[3] Luo Yifan. Research on Electricity Customer's Credit Rating and Anti-stealing Power Considering Distributed Generation. Shanghai, Shanghai Jiao Tong University, 2015.

[4] Streimikiene D, Ciegis R, and Grundey D. Energy Indicators for Sustainable Development in Baltic States. Renewable & Sustainable Energy Reviews, vol. 5, Dec 2007, pp. 877-893.

[5] Niu Xiaomei, and Zhang Yinling. Assessment of Power Customer Credit Risk Based on Analytic Hierarchy Process. Computer Simulation, vol. 5, 2011, pp. 333-336.

[6] Huang Yuansheng, and Shi Xiufen. Evaluation of Electricity Customers' Credit Risk based on Entropy Method and MADM. Technoeconomics & Management Research, vol. 2, 2010, pp. 24-27.

[7] Lu Jianchang, Jiang Wei, and Liu Na. Study on Credit Evaluation of Power Consumers Based on Improved Entropy. Journal of Electric Power, vol. 5, 2008, pp. 368-370.

Acknowledgment

This work is sponsored and supported by National Power Grid Corp science and technology project *Research Application of Key Technologies for Power Grid Planning, Investment and Purchasing Power Optimization* (Contract No. SGZJJY00SJJS1600031).

IOP Conf. Series: Earth and Environmental Science **127** (2018) 012012 doi:10.1088/1755-1315/127/1/012012

Application of Time-series Model to Predict Groundwater Quality Parameters for Agriculture: (Plain Mehran Case Study)

Mir Mehrdad Mirsanjari[1], Fatemeh Mohammadyari[2]

[1]Department Environmental Scieneces, Faculty of Natural Resources and Environment, Malayer University, Malayer, Iran
[2]PhD Student of Evaluation and land use Planning, Department Environmental Scieneces, Faculty of Natural Resources and Environment, Malayer University, Malayer, Iran

corresponding author's e-mail : mehrdadmirsanjari@yahoo.com

ABSTRACT. Underground water is regarded as considerable water source which is mainly available in arid and semi arid with deficient surface water source. Forecasting of hydrological variables are suitable tools in water resources management. On the other hand, time series concepts is considered efficient means in forecasting process of water management. In this study the data including qualitative parameters (electrical conductivity and sodium adsorption ratio) of 17 underground water wells in Mehran Plain has been used to model the trend of parameters change over time. Using determined model, the qualitative parameters of groundwater is predicted for the next seven years. Data from 2003 to 2016 has been collected and were fitted by AR, MA, ARMA, ARIMA and SARIMA models. Afterward, the best model is determined using information criterion or Akaike (AIC) and correlation coefficient. After modeling parameters, the map of agricultural land use in 2016 and 2023 were generated and the changes between these years were studied. Based on the results, the average of predicted SAR (Sodium Adsorption Rate) in all wells in the year 2023 will increase compared to 2016. EC (Electrical Conductivity) average in the ninth and fifteenth holes and decreases in other wells will be increased. The results indicate that the quality of groundwater for Agriculture Plain Mehran will decline in seven years.

1. Introduction

In terms of water resources and reduced water quality, water challenge with effects on agricultural production, energy conservation, as well as environmental protection has become a global issue [1]. Groundwater which is an important part of the reproducible water ecosystems most often existed in arid and semi arid areas where surface water are very low [2]. Ground-water management is more difficult than surface water resource management; therefore, there is a need to seek for reasonable and cost-effective procedures to determine the status of these waters [3]. As a result, planning for the optimal use of water resources with the aim of achieving sustainable development is of utmost importance. a time series is defined as a sequence of observations ordered in time. The basic premise in the analysis of time series forecasting future values based on historical observations studied variables [4]. These models are widely used in hydrology and water quality parameters [5]. There are some researchers who have conducted different predictions by using one time-series model [6, 7] and other scholars who have employed several time-series models [8, 9]. Time series models in different fields cuusnt the attention researchers. such as air pollution [5], changes in groundwater levels [10] and water quality [11]. The combination of time-series models including AR, MA, ARMA, ARIMA and SARIMA to forecast ground-water quality parameters was provided by [12] (2014); but the combination of these models to predict water quality parameters has not been reported yet. Therefore; with respect to the conducted research studies and the ability of time-series techniques, this study is to examine the status of ground-water quality through different models of time series design by using 15 years of time series data of ground-water wells for agriculture in Mehran Plain, and eventually predict the status of ground-water quality parameters for the next 7 years (2021) via the best model obtained.

2. Materials and methods

2.1 Area of the Study

The study area (longitude: 46°05′to 46°15′E, latitude: 32°59′to 33°37′N) is located in Mehran plain, Ilam province, Iran. The Mehran plain has an area of 2391.9 km2. (Figure 1).

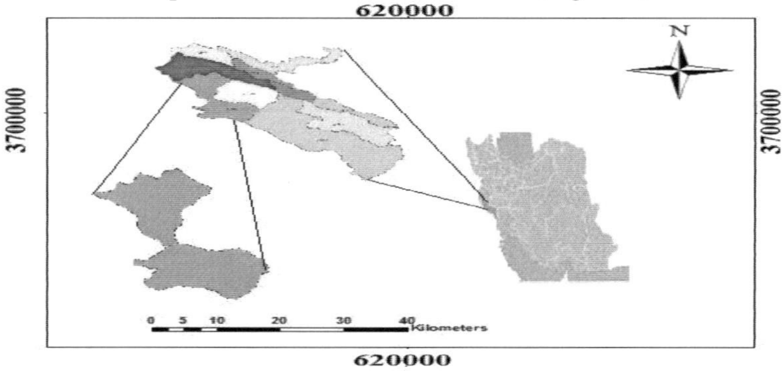

Figure 1. Location of the study region.

2.2 Methodology

Using data extracted from ground-water quality parameters for 17 wells in 2001 to 2015 in Mehran Plain, the variation process in these parameters were modeled and then the State of agriculture Plain were predicted for the next seven years via the selected model. Effective parameters in agriculture according to classification by Wilcox (1995), shown in Table 1 and Table 2 is was used (electrical conductivity and sodium adsorption ratio). The data used were monthly data and continuous measurements; therefore, the time series under investigation was of continuous type. Drawing time series data is the first step in the analysis of time series design. The goal is to determine the presence or absence of the process in the data. In the next step, the components of the process in time series are determined and then they are removed in order to make the data static. After that the stativity of the data is examined, the appropriate model is fitted to the data to identify the best model and accordingly make the prediction. The third step in the analysis of time series is the investigation into the normality of the data obtained from the prediction of the model; so that the Kolomogrov-Smirnov test was used to assess the normalization of data in this study. The models used in this study included AR, MA, ARMA, ARIMA and SARIMA. based on the models used in this study, there is no need to determine the equation of the line of goodness of fit for the data and the elimination of the process, and the models used remove the deletion process on the data by themselves. In the ARIMA and SARIMA models, the seasonal data status is also removed through differentiation. Therefore, in this study, R software was employed to determine the best time series model and ultimately forecast the data required by the selected model. Out of the 64 data quality parameters, 28 were simulated and set aside for the calibration of the model. To describe time dependence in the structure of a time series model, autocorrelation functions (ACF) and partial autocorrelation (PACF) were investigated. After the determination of goodness of fit for the model, detection of the accuracy of the selected pattern is essential. According to the charts of ACF and PACF functions, the best models were determined via information criterion such as Akaike (AIC) and coefficient of determination (R2). AIC test is used to compare different models of ARMA (p, q) and it is calculated as follows (Mirzavand and Ghazavi, 2014):

$$\text{AIC (k)= nLn (MSE)+2K} \tag{1}$$

Where n is the number of data points (for calibration), and k is the number of free parameters used in models. MSE stands for mean square error. Usually, the preferred model gives a higher R2 or the smallest value of AIC. After validation of the best model fitted on time series, to prediction future of observations are done. In the forecast process, the current time period is shown by t and t+τ represents the prediction for the time period of t+τ at the time of t. The forecast is made through considering the mean at the origin of t from the model written at the time of t+τ. In general, prediction is provided for the time period of t+τ-1, ..., t+2+t+1. In this method, xt+j which has occurred at the time of t is replaced with the predictions of ẍt+j (t), and εt+j that have not occurred at the time of t are substituted with zero. As well as, εt+j that have not occurred are replaced with a single-period prediction error of e1(t-j)=xt-j-ẍt-j(t-j-1) (Erdem and Shi, 2011). After selecting the best model and predicting values of quality parameters in 2021, the changes between actual (2015) and modeled year (2021) was. For this purpose, the experimental variogram of SAR and EC parameters in the first two years, 2015 and 2021 are calculated separately and the best model was selected for each parameter. After drawing and modeling variograms, we used interpolation technique with kriging in Arc GIS 10. Kriging interpolation is considered as an advanced method for the topical trends [13]. Finally, after obtaining the distribution of EC and SAR maps every two years, agricultural zoning map were generated and possible changes in the year 2021 were discussed.

2.3 Time Series Models

The auto-regressive model (AR) (p) as Eq. (2) can be expressed:

$$z_t = \varphi_1\, z_{t-1} + \varphi_2\, z_{t-2} + \ldots\ldots.\ \varphi_p\, z_{t-p} + a_t \tag{2}$$

Where φ1, φ2 and φp are coefficients and model parameters and The random term of the data follows normal distribution with a zero mean. The model MA can be expressed as Eq. (3):

$$z_t = \theta_1\, a_{t-1} + \theta_2\, a_{t-2} + \ldots\ldots.\ \theta_q\, a_{t-q} + a_t \tag{3}$$

Where θ1, θ2 and θq are coefficient and model parameters and at is a random term of the data that follows by normal distribution with a zero mean [14]. ARMA (p, q) model also can be obtained from the following equation (4):

$$Y_t = \delta + \sum_{i=1}^{p} \emptyset_i y_{t-1} + \sum_{j=1}^{q} \varphi_j e_{t-j} + e_t \tag{4}$$

Where δ is the constant term of the ARMA model, ϕi indicates the ith auto-regressive coefficient, φ_j is the jth moving average coefficient, et shows the error term at time period t, and Yt refers the value of groundwater level observed or forecasted at time period t [15]. ARIMA models originated from the combination of autoregressive models (AR) and moving average models (MA). ARIMA fits a Box-Jenkins ARIMA model to a time series [6]. In ARIMA, the future value of a variable is assumed to be a linear function of several past observations and random errors. A SARIMA model can be explained as ARIMA (p, d, q) (P, D, Q)s, where (p, d, q) is the non-seasonal part of the model and (P, D, Q)s is the seasonal part of the model in which p is the order of non-seasonal autoregression, d is the number of regular differencing, q is the order of non-seasonal MA, P is the order of seasonal auto-regression, D is the number of seasonal differencing, Q is the order of seasonal MA, and s is the length of the season [11].

Table 1. Classification criteria in terms of agricultural water

water quality	EC	Class	SAR	Class
Excellent	<250	C_1	<10	$\mathbf{S_1}$
Good	250-750	C_2	10-18	$\mathbf{S_2}$
Intermediate	750-2250	C_3	18-26	$\mathbf{S_3}$
Unsuitable	>2250	C_4	>26	$\mathbf{S_4}$

Table 2. Various categories and types of water quality classification based Wilcox [16]

Row	Water Category	Quality of water for agriculture
1	C1S1	Sweet-for Agricultural quite harmless
2	C2S1,C2S2,C1S2	Little salty-for Agricultural almost perfect
3	C3S3,C3S2,C3S1,C2S3,C1S3	Salted-for agriculture by applying the necessary measures
4	C4S1,C4S2,C4S3,C4S4,C3S4,C2S4,C1S4	Very Salt harmful for agriculture

3. Result and Discussion

12 models on groundwater parameters, the results, for four wells are shown in Table 3.

Table 3. Results of time series models in 4 Wells

Used models			Well 1			Well 2			Well 3			Well 4		
	parameter		Model coefficient	AIC	R2	Model coefficient	AIC	R2	Model coefficient	AIC	R2	Model coefficient	AIC	R2
AR(1)	EC	φ1	-0.4	509	0.837	-0.1	415	0.652	-0.4	342	0.817	-0.3	470	0.582
	Sar		-0.3	-22.5	0.729	-0.3	-9	0.726	-0.1	-57	0.837	-01	-34	0.643
AR(2)	EC	φ1 φ2	-0.6 -0.4	503	0.739	-0.2 -0.6	397	0.591	-0.6 -0.2	342	0.713	-0.4 -0.4	463	0.614
	Sar		-0.5 -0.3	-28.9	0.799	-0.5 -0.4	-14	0.672	-0.2 -0.5	-71	0.853	-0.3 -0.5	-46	0.610
MA(1)	EC	θ1	-1	490	0.842	-1	397	0.649	-0.7	336	0.816	-1	451	0.579
	Sar		-1	-40.7	0.732	-1	-24	0.726	-1	-71	0.834	-1	-52	0.645
MA(2)	EC	θ1 θ2	-1 0.4	490	0.843	-0.4 -0.5	396	0.646	-0.7 0.01	338	0.816	-1 0-1	452	0.585
	Sar		-1 0.1	-38.9	0.730	-0.9 -0.01	-22	0.726	-0.3 -0.6	-76	0.837	-0.6 -0.3	-51	0.645
ARMA (1,1)	EC	φ1 θ1	-0.1 -1	491	0.843	0.06 -1	399	0.650	-0.02 -0.7	338	0.816	-0.03 -0.9	453	0.580
	Sar		-0.05 -1	-38.9	0.731	0.06 -1	15	0.726	0.1 -0.9	-70	0.833	0.06 -1	-50	0.645
ARMA (1,2)	EC	φ1 θ1 θ2	-0.4 -0.5 -0.4	494	0.842	-0.3 -0.2 -0.7	396	0.640	-0.9 0.2 -0.7	337	0.804	-0.7 -0.1 -0.8	454	0.582
	Sar		-0.8 -0.08 -0.9	-37.1	0.733	-0.6 -0.2 -0.7	-21	0.732	-0.3 -0.1 -0.8	-76	0.842	-0.2 -0.4 -0.5	-50	0.645
ARMA (2,1)	EC	φ1 φ2 θ1	-0.1 -0.3 -1	489	0.849	0.1 -0.6 -1	381	0.635	0.02 0.05 -0.8	340	0.816	-0.06 -0.3 -0.9	449	0.576
	Sar		-0.08 -0.2 -1	-39.8	0.732	0.02 -0.2 -1	-22	0.735	0.2 -0.4 -1	-77	0.816	0.07 -0.4 -1	-58	0.647
ARMA (2,2)	EC	φ1 φ2 θ1 θ2	0.5 -0.1 -1 0.9	487	0.839	0.2 -0.7 -1 0.2	383	0.651	-0.6 0.3 -0.01 -0.9	338	0.803	0.6 -0.2 -1 0.9	448	0.573
	Sar		0.6 -0.1 -1 1	-40.5	0.733	0.8 -0.1 -1 -0.9	-21	0.713	-0.1 -0.3 -0.4 -0.5	-77	0.831	0.1 -0.4 -1 0.08	-56	0.647

To predict future fluctuations in groundwater quality series validation models for each time series was used for a period of 7 years. Figure 2 an example of real and simulated of that SAR and EC of the well third shows.

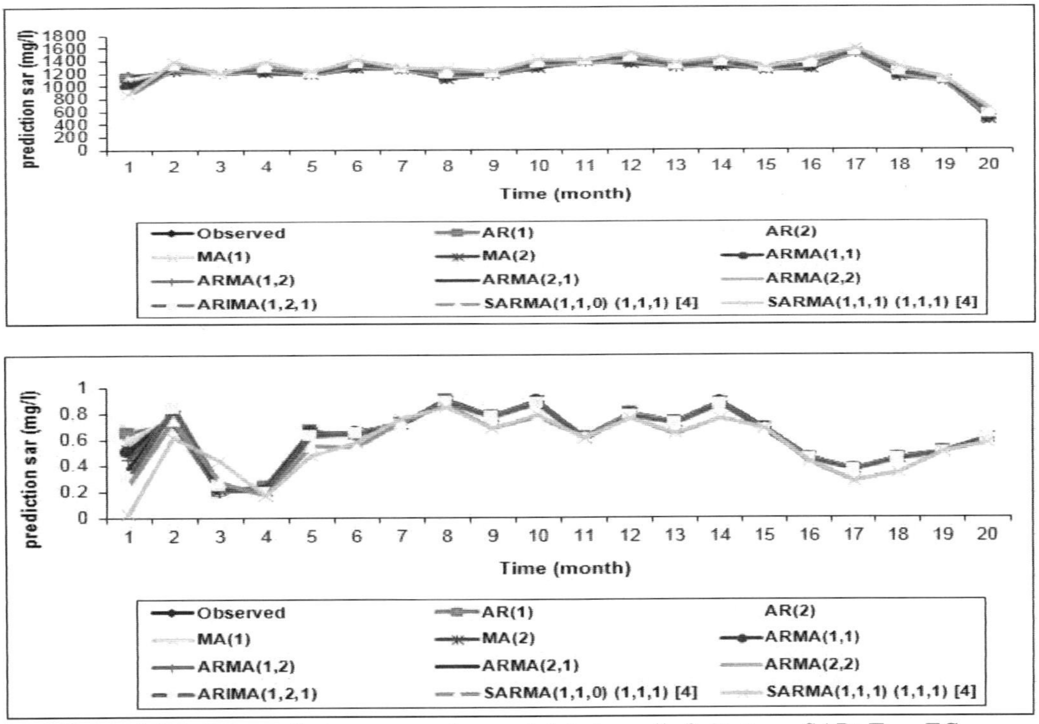

Figure 2. Models prediction versus observed data in wells 3, Bottem: SAR؛ Top: EC

Models were next fitted on parameters, the appropriate model to predict ground-water parameters were selected using Akaike and correlation coefficient indices. As shown in Figure 2 and Table 4, ARMA, AR and MA was selected as suitable model.

Tabel 4. wells final model Time Series

wells	Parametr	Models
1, 5, 9, 13, 16	EC	ARMA (2,1)
	SAR	AR (2)
2, 8, 11	EC	ARMA (2,2)
	SAR	MA (2)
3, 7, 12, 15, 17	EC	MA (1)
	SAR	AR (2)
4, 6, 10, 14	EC	AR (2)
	SAR	MA (2)

According to the results, the mean SAR predicted in the years 2022 to 2015 all wells will increase. EC average decreases in the ninth and fifteenth wells and will be increased in other wells. with the best chosen models by using software ARC GIS10.3, EC and SAR parameters were interpolated by kriging(Figure 3).

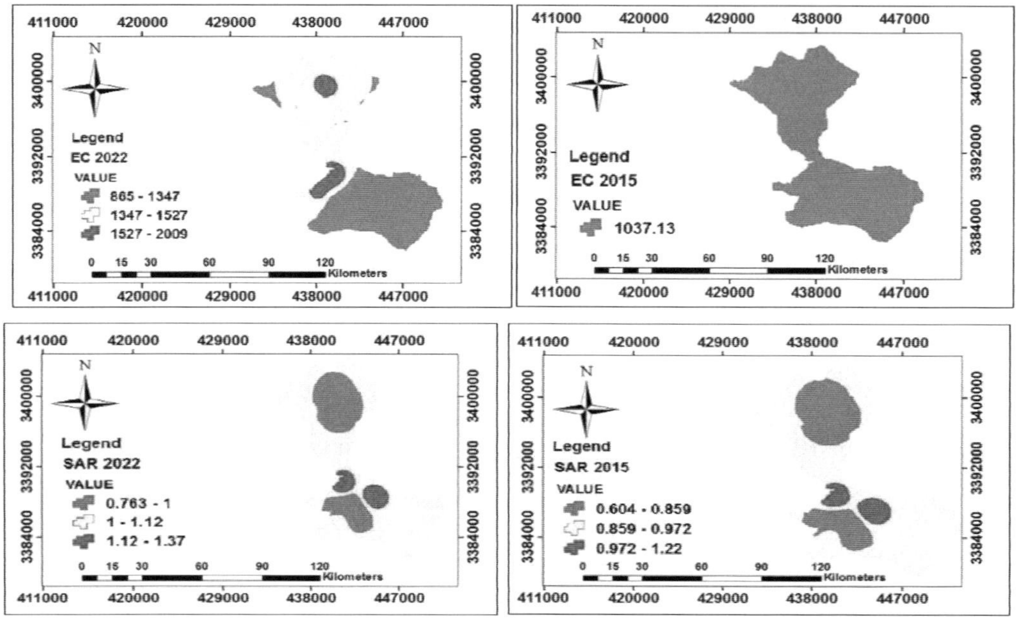

Figure 3. Distribution maps of groundwater water quality param

Finally **groundwater** quality in terms of agricultural zoning map was produced according to the Wilcox [16] are shown in Figure 4 and Table 5.

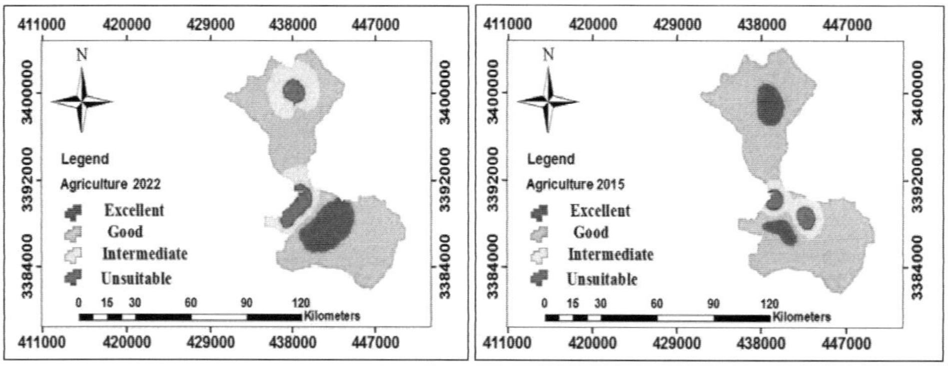

Figure 4. Zoning Agriculture maps of groundwater quality

Table 5. Area of the agricultural regional classes

	Year 2015		Year 2022	
Class	Area (km^2)	Area(%)	Area (km^2)	Area (%)
Excellent	206.61	9	279.7	12
Good	1938.3	80	1557.99	65
Intermediate	162.38	7	404.98	17
Unsuitable	84.63	4	149.13	6
Total	2391.9	100	2391.9	100

4. Conclusions

To predict the parameters of ground-water quality using monthly data, the quality parameters were analyzed into four sections including random, seasonal, trend and observed; and the random parameter was selected as the model. Additionally, five models with 12 sub-models were evaluated. AIC and R2 criteria for model selection were used to make more accurate predictions. Based on the results illustrated in Table 3 and according to AIC and R2 coefficient values for the different models examined, the final models of quality parameters were determined for each cluster (Table 4). In these models, R2 and AIC had respectively the maximum and the minimum values and the absolute value for the parameters of the selected models did not exceed 1. The results show that the time series models in the modeling of spatial and temporal variations of groundwater quality. The time-series models, four models of ARMA (1,1), ARMA (2,1), ARMA (2,2) and SARIMA (1,1,1) (1,1,1) [4] for ground-water quality parameters in mehran Plain were identified through the autocorrelation and partial correlation functions. The appropriateness of the model was confirmed by the analysis of the remaining fitted model. Normality test results showed normality of the data was predicted. From the results it is expected in 2022, the EC is higher than the standard for agriculture. The SAR value is higher than the standard in all wells for agriculture. The comparison of the data from model that is predicted by different methods and the original data are shown in Table 3 and Figure 2. These data demonstrate that the selected time series models have had a good performance in predicting the time series of ground-water quality parameters which are also consistent with the findings by [17, 9, 6]. The distribution maps of qualitative indicators modeled of the year 2022 reflects the deteriorating quality of groundwater resources in the future years. as indicated in Figure 3, EC value is appropriate in all parts of the plain in 2015 but it is gradually decreased and in the year 2022 a small part of the plain possesses relatively adequate value and the majority of it which lie on the floor would be undesirable. The SAR region in 2022 and a few changes in gradation, as compared to the year 2015 follow the trend of declining quality stems. The final map in 2015 due to higher agricultural area for the middle class and 4% of the area is unsuitable for agriculture. But in 2022 the extent of suitable area for agriculture will be reduced that reflects alarming circumstance of water quality for agricultural utilization (Table 5). Irregular exploitation of groundwater aquifers and unauthorized wells are the most important reasons of quality decrease in Mehran Plain. Protecting approach may include pressurized irrigation projects, recharge artificial wells and installing water meters on agriculture, unauthorized filling of wells and finally raising awareness. Furthermore, considering the studied area which is located in arid and semi-arid area and dealing with water scarcity, the results shows a clear picture of the groundwater quality parameters in future and can be used by professionals and planners.

References

[1] Wu D, Ekama GA, Wang HG, Wei L, Lu H, and Chui H 2014 Simultaneous nitrogen and phosphorus removal in the sulfur cycle-associated Enhanced Biological Phosphorus Removal (EBPR) process. *Water Research* **49**:251-264.

[2] Nas B, and Berktay A 2008 Groundwater quality mapping in urban groundwater using GIS. *Environ monit Assess.* **160 (1–4)**: 215–227

[3] Hajihashemi Jazy MR, Atashgahi M, and Hamidian AH 2011 Spatial estimation of groundwater quality factors using geostatistical methods (case study: Golpayegan plain). *Journal of Natural Environmental, Iranian Journal of Natural Resources*, **4**: 347-357

[4] Taneja K, Ahmad sh, Ahmad K, and Attri SD 2016 Time series analysis of aerosol optical depth over New Delhi using BoxeJenkins ARIMA modeling approach, *Atmospheric Pollution Research*, **3**: 1-12

[5] Wang H R, Wang C, Lin X, and Kang J 2014 An improved ARIMA model for precipitation simulations. *Nonlinear Process. Geophys.* **21**: 1159-1168.

[6] Shirmohammadi B, Vafakhah M, Moosavi V, and Moghaddamnia A 2013 Application of several data-driven techniques for predicting groundwater level. *Water Resour Manag* **27**:419–432.

[7] Adamowski J, and Chan HF 2011 Awavelet neural network conjunc tion model for groundwater level forecasting. *Hydrol* **407(1–4)**:28–40. doi:10.1016/j.jhydrol.2011.06.013

[8] Parmar KS, and Bhardwaj R 2014 Water quality management using statistical analysis and time-series prediction model. *Appl Water Sci.* **4**: 425–434.

[9] Malekian A, Chobin B, Sajedi Hossini F, and Rahmati A 2015. Prediction water table using time series and Fuzzy Inference System - Adaptive Neural, *Journal of Soil and Water Research Iran*. **45**: 19-48.
[10] Panda D K, and Kumar A 2011 Evaluation of an over-used costal aquifer (Orissa, India) using statistical approaches. *Hydrol. Sci. Jour.*, **56**: 486-497.
[11] Faruk D 2010, A hybrid neural network and ARIMA model for water quality time series prediction. *Eng Appl Artif Intell* **23(4)**:586–594. doi:10.1016/j.engappai.2009.09.015
[12] Mirzavand M, and Ghazavi R 2014 Stochastic Modelling Technique for Groundwater Level Forecasting in an Arid Environment Using Time Series Methods, *journal Water Resources Management*. Doi: 10.1007/s11269-014-0875-9 Online ISSN 1573-1650.
[13] Ghahrevardi Tally M 2005 *In the three-dimensional geographic information system ArcGIS*, Publications Teacher Training University Jahad.
[14] Hannan EJ 1971 *Multiple time series*. Wiley, New York
[15] Erdem E, and Shi J 2011 ARMA based approaches for forecasting the tuple of wind speed and direction. *Appl Energy* **84(2)**:1405–1414. doi:10.1016/j.apenergy.2010.10.031
[16] Wilcox L V, 1955 *Classification and use of irrigation*
[17] Mirzaee S, Chitsazan M, CHinipardaz R, and Samady H 2010 Forecast plain SHAHREKORD using time-series models and examine ways to improve, *The first regional conference on optimal utilization of water resources and river basins Karon*. **46**: 1-8.

IOP Conf. Series: Earth and Environmental Science **127** (2018) 012013 doi:10.1088/1755-1315/127/1/012013

Let a sewage plant running smart

Shan-Shan Yang[1], Ji-Wei Pang[1,*], Xiao-Man Jin[1], Zhong-Yang Wu[1], Xiao-Yin Yang[1], Wan-Qian Guo[1], Zhi-Qing Zhao[1], Nan-Qi Ren[1,*]

1 Harbin Institute of Technology, Harbin 150090, PR China

*Corresponding authors' E-mail addresses: crabpjw@126.com (Ji-Wei Pang); rnq@hit.edu.cn (Nan-Qi Ren)

Abstract. Out-dated technical equipment, occlusive information communication, inadequate sanitation, low management level and some irrational distribution structures in the existing sewage plants bring about lower sewage treatment efficiency and poorer water quality, thereby permanently harming human health and severely damaging the environment. With the rapid development of scientific-technological progress and the vigorous support of the entire international community, the existing sewage plants call for more and more intelligent operation and management in the future. This review for the first time proposes the novel concept of the "smart" sewage plant, and gives a through interpretation of its special functions and attributes. We envision that the future smart sewage plant will became an "ambient intelligence" in all aspects in the sewage plants.

1. Why Do We Need a Smart Sewage Plant?

Water resource is critical for sustainable development. With the acceleration in the urbanization process, the increasing generation of domestic and industrial wastewater has already attracted much more public attention and awareness in many countries in the world [1]. According to the United Nations World Water Development Report, by 2050, there will be at least one in four people live in a chronic or recurring freshwater shortage country [2]. Thus in the coming years, people are going to face more and more serious water resource challenges.

The United Nations has always been committed to water programs and planning, e.g. resolution A/RES/47/193 of 22 December 1992 (Observance of World Day for Water) [3], resolution A/RES/55/196 of 20 December 2000 (International Year of Freshwater, 2003) [4], resolution A/RES/58/217 of 23 December 2003 proclaimed the period 2005-2015 International Decade for Action "Water for Life" [2]. In the report of the Second Committee (resolution A/RES/65/154), General Assembly declared 2013 the International Year of Water Cooperation [5]. For the Year, the U.N. announced its strongest calling for "all Member States to take advantage of the Year to promote actions at all levels, including through international cooperation, as appropriate, aimed at the achievement of the internationally agreed water-related goals, as well as to increase awareness of their importance" [5]. How to solve the water resource problem, has become a vital challenge that each country in the world is now facing and in urgent of solving.

Water resource shortage problems that are particularly challenging in China, have already become an important factor restricting China's social and economic development [6]. On 19 April 2012, the General Office of the Chinese State Council issued a notice of 12th Five-Year Plan on "The Urban

Sewage Treatment and Recycling Facility Construction Plan" [7]. The "Notice" stressed that during the 12th Five-Year Plan, the government's total investment on construction of sewage treatment and recycling facilities will amount to CNY 430 billion (U.S. $ 70.24 billion), which increased by 22.79% of total investment compared with that in the 11th Five-Year Plan (U.S. $ 54.32 billion).

Although Chinese government ramped up the capital investments, this "Notice" focuses mainly on facilities construction and sewage treatment capacity, still not sufficient. According to the latest report from China's water network (available at http://www.h2o-china.com/), there are about 32% of the existing sewage plants were built and put into operation before 2006, in which the sewage treatment capacity is account for 50% of the total capacity today. Among all of these sewage plants, there are only 12% can meet class A Integrated Wastewater Discharge Standard (GB8978-1996), however, these plants only occupied less than 7% of the total sewage treatment capacity. While the number of sewage plants meeting class B (GB8978-1996) accounted for 54%, and still 34% of the sewage plants discharge is below class B (GB8978-1996). The fundamental reasons for lower sewage treatment capability and quality in traditional sewage treatment plants may due to outdated technical equipment, irrational industrial structure, occlusive information communication network, high energy consumption, as well as low level of management. With more and more ambient intelligence improved our daily life [8], it is strongly needed to make our sewage plants more smart and intelligent.

As the 12th Five-Year Plan proposed that the development of sewage plants in China is entering the "Upgrading and Reconstruction Time", the traditional sewage plants will arouse increasing intelligent operation and management [9]. The technologically innovation with intelligence, automation, digitalization, flexibility, resilience, and sustainability operation and management will be strongly recommended in the future. The construction and extension of the new-generation sewage plants, which entitles "smart sewage plant", will forward new challenges to create a completely intelligent environment in the future sewage plants in China.

2. What is a Smart Sewage Plant?

The realization of "smart" for a future sewage plant needs four essentials (Figure 1):

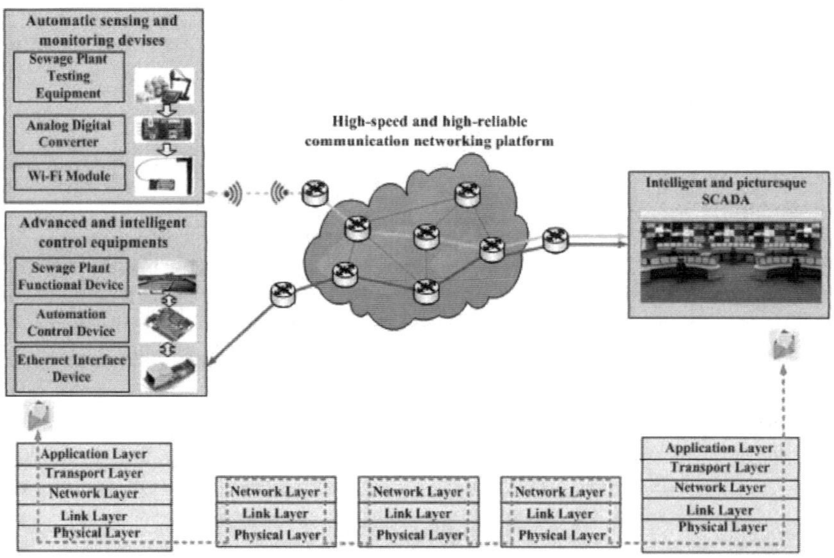

Figure 1. The messages transmitted in the smart sewage plant are based on typical Ethernet 5-layer protocol architecture. The communication between the field equipment and SCADA can be realized through Wi-Fi and fixed optical fiber.

- ***Intelligent and picturesque supervisory control and data acquisition (SCADA).*** Intelligent and picturesque SCADA could provide deeper understanding of the overall sewage plant performances. All the intelligent decisions are automatically made by SCADA, and subsequently sent to the various control terminals. Via a local or remote control operations, the network communication platform will expand to the SCADA to make the collected data readily and conveniently available.

- ✓ ***Advanced and intelligent control equipment.*** Advanced and intelligent control equipment are the key essential in determining the behavior of the smart sewage plant. With the participation of more advanced and intelligent control equipment, a smart sewage plant is expected to detect, analyze and respond smartly to the changes (or the shocks) in the operation or to predict the possibility of failures at the earliest without human intervention.
- ✓ ***Automatic sensing and monitoring devices.*** Automatic sensing and monitoring devices embedded in the control equipment efficiently guarantee the intelligent decision-making strategy. These devices in the smart sewage plant play the fundamental role in enabling the transformation of data into straightforward information. With the introduction of advanced sensing and monitoring devices, the smart sewage plant can evaluate the health of the sewage plant operation in real time, enabling a rapid smart diagnosis and supporting the optimized control strategy.
- ✓ ***Intelligent communication networking.*** Sophisticated wired and wireless communication networking allow a "plug-and-play" environment, which these can ensure the high-speed and high-reliable transmission of controlling and sensing information from intelligence terminal devices to smart control center [10]. We can envision that in the future, some portable interfaces, for instance, iPad, iPhone, and Android mobile phones can be exploit by the operators to carry out simple and convenient check, to control the operation state of the sewage plant in a wireless way.

3. What Makes a Smart Sewage Plant Smart?

Exactly what is a smart sewage plant? Actually, the smart sewage plant is not defined by how advanced equipment it integrates but rather by what it can smartly cooperate. The running for a future smart sewage plant is to provide the means to implement scientific decision-making by smart control algorithms. It can provide an optimal sewage treatment strategy smartly itself.

3.1. Real-Time Modelling

For the future smart sewage plant, real-time modelling is a novel concept. Through the collection of the real-time information and real-time monitoring data from the whole sewage treatment plant, real-time modelling gives the accurate analysis and evaluation of its present operating state and builds the corresponding regulation and control model. Real-time modelling, as one of the core functions, could give a variety of strategies to optimize the sewage treatment process.

3.2. Intelligent Decision-Making

The new generation smart sewage plant will reduce the number of workforce and require highly educated and well-trained managers who can keep skills fresh and up-to-date in response to the growing requirement of the smart sewage plant. The future sewage plant is expected to automatically detect the possible changes and faults, continuously check and report the operating state of a sewage plant, and give the intelligent decision-making and self-healing based on real-time modelling.

3.3. Stable and Fast Security Analysis

Stability and Security are the critical issues in the future smart sewage plant. An intelligent and strong smart sewage plant must guarantee a stable and fast security analysis for the sewage plant operation. The future smart sewage plant will rely extensively on the use of the safe and stable online monitoring and real-time analysis technologies. The application of real-time analysis in the future smart sewage plant would provide more scientific protection on human, production, operation, fire, environment security and safety.

3.4. Self-Healing/Self-Correction System

Self-healing/self-correction system is a high performance system which is integrated into the future smart sewage plant. It is expected that the smart sewage plant will detect, analyze and respond automatically to provide a set of optimal self-healing and self-correction control strategies according to the real-time information feedback without human intervention.

3.5. Wireless Communication Networking

Advancement in communication networking technology is another significant aspect in realizing a smart sewage plant. In the future smart sewage plant, wide application of the wireless communication networking would enable us to transfer the collected information over both short-range and long-range communications. We can envision that in the future, some portable interfaces, for instance, iPhone, and Android mobile phones can be exploit by the operators to carry out simple and convenient check.

3.6. Superior Effluent Quality

Effluent quality is another significant element to evaluate the operation performance of a sewage plant, as any changes (e.g. influent quality fluctuation, temperature variation, mechanical fault and etc.) can in turn do harm to the environment. Unlike the existing sewage plants, the future smart sewage plant provides many intelligent and powerful functions and allows for the visualization, flexibility, interoperability, resilience, and sustainability of operation [9]. Superior effluent quality is hence confident, and helps enhance the environment protection.

3.7. Humanized and Visualized Interoperability platform

A humanized and graphical visualization interoperability platform will offer more convenient and simpler operation to the sewage plant operators. As it can display the function keys and modules on the SCADA, it is more friendly and useful for this interoperability platform to provide simple, well defined, humanized, and graphical operation. Thus the operators could promptly give operation instructions and actionable commands.

4. What Are the Perspectives for a Smart Sewage Plant?

Above all, a future smart sewage plant can intelligently provide the corresponding control decision; implementation program and emergency solution. As the smart sewage plant is featured by automated information, we depict the blueprint of a future smart sewage plant here (Figure 2).

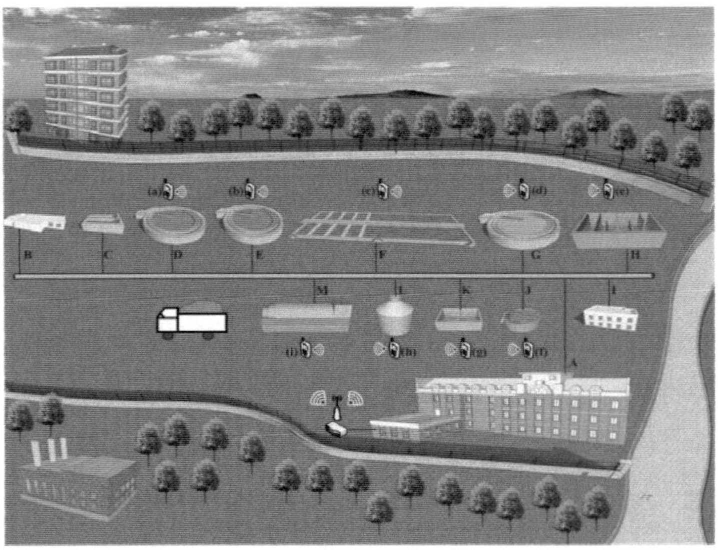

Figure 2. The scope of a smart sewage plant extends over all the aspects of the sewage treatment processes.

5. What Are the Main Roadblocks for a Smart Sewage Plant?

Implementation of the smart sewage plant involves many formidable obstacles and anticipative challenges. One critical challenge of replacing the existing sewage plant with a smart sewage plant is the demand for a unified communication protocol. As is mentioned above, the realization of "smart" for sewage treatment plant needs advanced and intelligent control equipment and automatic sensing and monitoring devices. Different equipment and devises purchase directly from different manufacturers have their own communication protocols; a common understanding is the ability to access any data from anywhere. However, an unify communication protocol standard plays an important role in a smart sewage plant by providing an unified way of accessing and exchanging data related to equipment and devices. For a long-term hard task, the need for a unified communication protocol is therefore obvious and imperative for the efforts of researchers.

Another challenge for the future smart sewage plant is to set a unified operational platform in the smart control centre. With the introduction of unified operational platform and standardized operation manual, a future smart sewage plant will not only realize the resources sharing between different sewage plants, but also can achieve a wide range of centralized monitoring and management.

6. Conclusion

Even as we confront the obstacles and challenges before us, overcoming the obstacles and challenges are inevitable and technical-enabled. In this review, we have defined a future smart sewage plant, its further development is indispensable for human quality living and long-term development. When looking beyond the horizon, we envision a brilliant and better future for a smart sewage plant.

7. References

[1] Liu, J. G.; Yang, W. Water Sustainability for China and Beyond. Science 2008, 337, 649-650.
[2] United Nations General Assembly. International Decade for Action, "Water for Life", 2005-2015 A/RES/58/217 (Distr.: General, 9 February 2004).
[3] United Nations General Assembly. Observance of World Day for Water A/RES/47/193 (93rd plenary meeting, 22 December 1992).
[4] United Nations General Assembly, 2003. International Year of Freshwater. A/RES/55/196 (Distr.: General, 1 February 2001).
[5] United Nations General Assembly, 2013. International Year of Water Cooperation. A/RES/65/154 (Distr.: General, 11 February 2011).
[6] Liu, X. F.; Yu, X. F.; Yu, K. The Current Situation and Sustainable Development of Water Resources in China. Procedia Engineering 2012, 28, 522-526.
[7] General Office of Chinese State Council, 12th Five-Year Plan on "The Urban Sewage Treatment and Recycling Facility Construction Plan" [in Chinese], No. 24, 2012. (available at http://www.gov.cn/zwgk/2012-05/04/content_2129670.htm.).
[8] O'Grady, M.; O'Hare, G. How Smart Is Your City? Science 2012, 335, 1581.
[9] Qin, X. S.; Gao, F.; Chen, G. H. Wastewater quality monitoring system using sensor fusion and machine learning techniques. Water Res. 2012, 46 (4), 1133-1144.
[10] Suh, K.; Smith, T.; Linho, M. Leveraging Socially Networked Mobile ICT Platforms for the Last-Mile Delivery Problem. Environ. Sci. Technol. 2012, 46 (17), 9481-9490.

Acknowledgments

The authors gratefully acknowledge the financial support by the National Nature Science Foundation of China (Grant No. 51708154), the Special Financial Grant from the China Postdoctoral Science Foundation (Grant No. 2016T90302), the class General Financial Grant from the China Postdoctoral Science Foundation (Grant No. 2016M601434), and the State Key Laboratory of Urban Water Resource and Environment (Grant No. HC201621-01).

IOP Conf. Series: Earth and Environmental Science **127** (2018) 012014 doi:10.1088/1755-1315/127/1/012014

Stormwater Management in Urban Areas of South Korea

S A Memon[1,2], O S Raja[2], B Kandhro[1], I Salim[2], C-H Lee[2]

1Institute of Environmental Engineering & Management, Mehran University of Engineering & Technology, Jamshoro, Sindh, Pakistan
2Department of Environmental Engineering and Energy, Myongji University, Yongin South Korea

sheerazahmed.memon@faculty.muet.edu.pk

Abstract. In early start of monitoring, a pathway for high runoff volumes and peak flows during rainfall period towards downstream of a waterbody was provided from storm sewer network, but later on it was realized to deal with stormwater quantity and quality to develop new approaches and management techniques. In early 90's NPS pollution issue was highlighted in korea, but only limited studies were conceded out up to the year 2000, however reasonably huge numbers of studies were directed for environmental science. After the recognition of NPS, Ministry of Environment in 1998 has introduced NPS as a major contributor in total maximum daily load management system (TPLMS) and waterbodies impairment, which is one of the guidelines of widespread water improvement strategies for main rivers. It contains a number of agendas that intention is to improve, maintain or restore the water quality in national water systems. It can be potted that stormwater management has evolved during the decades as of understanding with its impacts and it has been evolved from focusing on flood control to now incorporating control for volume, erosion and water quality, which is theoretically based on a watershed concept.

1. Introduction

In Korea, many treatment methods and restoration programs improved the water quality; however, according to National Environmental Policy Act of Korea it could not meet the established standards. Stormwater runoff pollution is a complex monitoring system that needs to measure during rainfall as well as in flow conditions. The monitoring of stormwater runoff from pre-during post construction phases have extreme significance in a newly urban development project. The onsite land disturbing activities such as cutting of trees, clearing, digging, and grading can significantly impair and pollute the water quality from stormwater pollution sources such as industrial, forests, agriculture and an uncontrolled and unmanaged construction industry. The management of these activities are difficult due to construction activities but it can be mitigating through good planning and best management practices because its impact on water quality is transient. During development of urbanization, transformation of land from pervious to impervious cover is an essential part but receiving water bodies have significant impact due to increase in anthropogenic activities and rapid urban growth [1]. The increased percentage of impervious surfaces such as roofs and roads by catchment results of urbanization. Subsequently the water quantity and water quality impact was increased during rainfall as illustrated in Figure 1 & Figure 2 [2]. This study provides the descriptions of the processes used to plan and monitor a successful stormwater runoff study from urban development sites. A large scale watershed monitoring such as river basins do not address directly by the approaches and guidelines of

IOP Conf. Series: Earth and Environmental Science **127** (2018) 012014 doi:10.1088/1755-1315/127/1/012014

this study but it emphases on specific land use variety as a particular place monitoring to reduce the pollution at the source, prevent, and control.

2. Methodology

Monitoring strategy of the study explain five points of why, what, how, when and where to monitor for stormwater source pollution.

2.1 Monitoring plan

A monitoring strategy developed before the start of monitoring project described by monitoring plan. In previous studies, the development and implementation of monitoring study was suggested by a three phase based approach [3]. The scope of specific monitoring study, determination of local monitoring needs and implementation of monitoring study include in three phase-based approach. There are the chances to lose potential for collecting sufficient data for defining the goal and objectives of monitoring project in its first step. In this study, monitoring plan improved every year according to last year monitoring plan considerations during 2011-2013 period.

A baseline monitoring, management facilities performance monitoring and wet weather event based monitoring are three different ways for stormwater study of monitoring types. Baseline monitoring describes to evaluate the impacts level by determine the reference or pristine and quality of background stormwater in receiving waterbodies earlier to following a development project or urban development. The post development monitoring can utilize the data as reference data by conducting baseline monitoring in catchment before the change of land use or construction. The stormwater treatment facilities performance and efficiency evaluated by monitoring of stormwater treatment facility. The management practices perceive and assess the performance level. The detection of any negative impact on pollutant concentrations were conducted in wet and dry conditions, which provided useful information and data evaluation. During runoff producing rainfall events wet weather event monitoring demonstrated the examination and stormwater evaluation. In this case, of monitoring, the key objective is to accomplish site-specific land use and hydrological settings of a catchment during and subsequent storm events.

2.2 Monitoring site selection

The visual inspections through regular field visits are a method of determining the monitoring site selection throughout pre-storm and post-storm conditions. During stormwater event the source area for the exposed pollutants were determined. It pacts with erosion, movement of soil, and checking of clear runoff, waste and chemicals that are exposed. During onsite field visits non-visible pollutants cannot be investigated in many cases and so stormwater analysis and sampling is appropriate and important.
Descriptive stormwater pollution parameters from single land use evaluated by a specific study sites involve the samples collection to quantify the effects of certain land uses, and to evaluate Best Management Practices performance. A homogenous land cover and drains out sufficient runoff is defined by this type of site includes source area to the stormwater drainage system. It is selected to capture, control, and treat the pollution at the target source based on provided that a system.

The precise study objectives such as if a monitoring study was directed for highway runoff then several features containing normal daily traffic and others were deliberated for selection of representative sites. In addition, the satisfactory data is necessary for statistical analysis provided by monitoring of stormwater events and number of sites [4]. In previous studies, similar criteria were followed as shown in Table 1.

The one of significant aspect in stormwater monitoring strategy is an exact location of sampling site after site selection and watershed has not guideline only regarding spatial features but furthermore in favour within a water body at a known position. The site for sample collection should be recognized where it should signify the total runoff from catchment area and the runoff from the site is discharged towards the receiving stream. The selection of any representative site must follow several considerations but individual local site-specific characteristics were represented from monitoring site [5].

Table 1. Specific examples of source area monitoring applications in previous studies

Examples of Source area monitoring studies	
Study Objective	**References**
Study of non-point source pollution in a mixed forest watershed	[6]
Determine the pollutant loading from roofs	[7]
Pollutant Washoff and Loading from Parking Lots	[8]
Characteristics of Stormwater Runoff in Different Land Use Areas	[9]
Long-Term Monitoring of Infiltration Trench for Nonpoint Source Pollution Control	[10]
Performance of a California Stormwater pond	[11]

2.3 Acceptable storm event criteria

The potential features for the selection of specific stormwater sampling and limits the project monitoring in money and personnel, to monitor every storm event described by storm event sampling criteria.

2.3.1 Minimum numbers of events. There is no general defined criterion for determining the number of rainfall runoff events but it depends on statistical evaluation. It should be adequate to provide realistic seasonal variations throughout the year and it should variate in size from low to high rainfall range. In previous studies seven events/year according to US EPA, 12-35 events per application US [12] and 6-8 events/year (north Oakville) were selected.

2.3.2 Antecedent dry period. It defines minimum dry days prior to the storm event. It is directly proportional to the pollutants accumulation and therefore causes more pollution during stormwater event. According to US-EPA with a minimum of 72-hours dry period preceding the storm event is required to monitor.

2.3.3 Stormwater Duration. It defines the adequate runoff duration of stormwater at beginning and end. According to site-specific characteristics consideration this feature mainly depends on rainfall-runoff lag time. For a single site trend characterization, it becomes difficult to estimate the duration of sampling needed. It can vary greatly depending on the purpose of the monitoring program. In previous study, 6 hours minimum stormwater duration was suggested [13].

*2.3.4 Rainfall Depth. I*t is the total depth of rainfall expected to occur over a given time period. It is considered from the period of attention prior to the time of concentration before the last sample being taken. It was observed from literature review that smallest depth is essential to create runoff for appropriate quantity of samples such as 0.15 inches Washington and 0.2mm Francey for urban monitoring sites [13, 14].

2.3.5 Average Rainfall Intensity. The energy supplied and storm event succeeds at the lowest rainfall intensity by pollutant removal from rainfall influences. It depends on site-specific characteristics and study objective.

IOP Conf. Series: Earth and Environmental Science **127** (2018) 012014 doi:10.1088/1755-1315/127/1/012014

2.4 Sampling Techniques

This is one of the perilous features in stormwater pollution characterization. In previous studies, two separate strategies have been recommended: time-proportional, time-weighted, or flow weighted or flow-proportional sampling and fixed frequency-sampling strategies. In present work, all monitoring sites considering time and flow throughout the stormwater event was sampled by manual grab sampling. Grab samples are discrete samples. During the stormwater events grab samples can be collected at any time and its flow change. Throughout the storm event the sampling time and the intervals between the samples is of great importance because of flushing and contaminants Washoff features relying on hydrological, site specific and topographical features. Another important aspect of any research is a sufficient number of samples collection and assurance of representative samples. It is suggested that the uncertainty in water quality measurements was reduced by using small sampling intervals.

3. Results

3.1 Monitoring site location

In present study, the criteria for monitoring sites selection was adapted based on pre-development catchment area, initial phase of trees cutting trees; continuous land disturbance activities phase (soil excavation, soil digging, bare land and others) from 1st year to 3rd year of the monitoring period.

First year objective was to monitor stormwater runoff from pre-development and starting phase of under development within same catchment area from Yongin city Korea. Second and third year of the study was involved in monitoring of watershed during continuous land disturbance construction site activities along with other catchment areas to monitor and compare the impact of construction area with other land uses. Figure 1 show all monitoring sites location year wise.

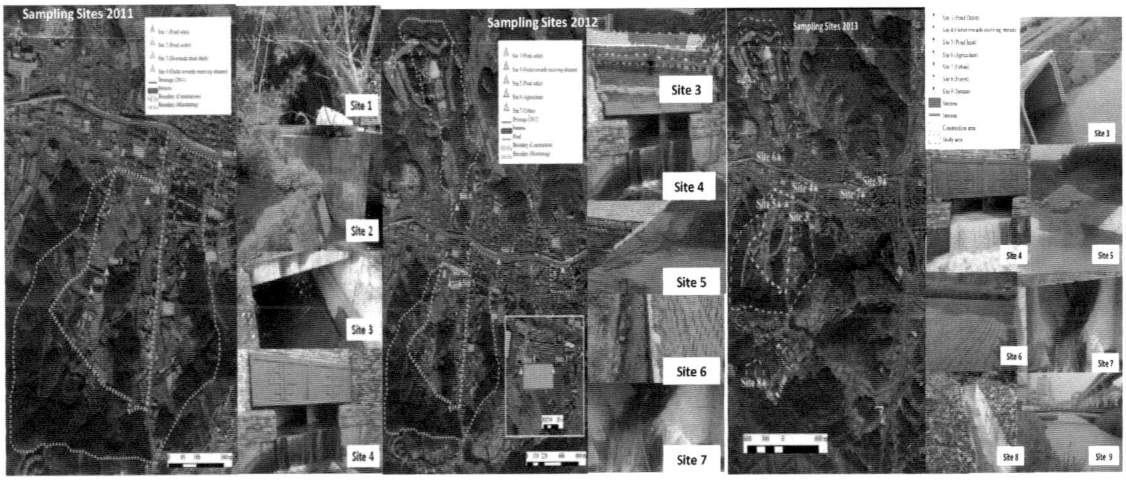

Figure 1. Monitoring sites location year wise

A specific land use type to a small watershed scale from 2011 to 2013 delivers monitoring sites selection criteria. During construction phase on-site difficulties of diverse flow patterns, unsafe conditions to select catchment area causes the scope of the study was extended and to investigate impact of construction site industry on small-scale watershed catchment.

After site selection, the one of the important aspects in stormwater strategy is exact location of sampling site and instruction not only about spatial features within a watershed but also in concern at a given location inside a water body. The runoff leaves the known catchment without interference of any other site identifies sample collection site. Throughout monitoring period, these locations should be fixed and stable without any alteration or modification in physical drainage patter. In present work, due to non-stable soil covers, grading, land disturbance activities, digging of soil, excavation and characteristics of discharge and non-representation of flow construction site location was not similar every year.

During field visits for site location, more than two inlets were found in sedimentation pond. Therefore, it was difficult to select one discharge point as monitoring site location. It is advisable and important to collect discharge of catchment area from one single pipe outlet or drainage channel, therefore multiple inlets from different land disturbing areas should be connected all together to represent as one single discharge point.

Practices during different phases of construction and development work were not exercised which caused polluted runoff entrance to road runoff, soil erosion and direct discharge of runoff to receiving water body without prior control practices as shown in figures. In this study it is suggested to install and fix temporary management practices such as sedimentation trap from small disturbed area, silt fences, storm drain inlets protection, according to land disturbance activities in construction phase.

3.2 Acceptable stormwater criteria

Precipitation is an important feature in stormwater monitoring to select representative events. In this study, it is categorized according to different rainfall classes reviewing past 10 years' rainfall data (<4mm, 4-10mm 10-30mm, 30-50mm, and >50mm) and therefore minimum of 8 to 10 representative stormwater events was selected. Result findings and literature review reveal that the lowest numbers of stormwater events are essential six times in a year. In case of antecedent dry period, more than one dry day period in 1st year and three or more than that were selected for remaining years. Stormwater duration was followed according to beginning of the runoff till it becomes clear due to unpredicted trend of construction site runoff. In general, minimum of six hours' duration was fixed because shorter duration can cause insufficient and poor representative samples. Urban site was observed with 1 mm minimum depth of rainfall to the total rainfall event, whereas, rainfall depth >4mm was obtained from stream sites and mix catchment that was monitored for the events. While the requirement for minimum rainfall depth was observed >10mm in case of construction, agriculture, and further pervious sites. Therefore, two different criteria were followed such as at least 1mm rainfall for impermeable cover area and 10mm for pervious area. Similarly, average rainfall intensity was fixed as > 1.5mm/hr for pervious sites and > 0.2mm/hr for impervious sites but it was observed that they depend on total rainfall and site-specific characteristics.

3.3 Stormwater monitoring

Discrete sampling was conducted in this study to reduce the sampling error and capture within storm variability. It is simple to manage therefore recommended to estimate pollutant loads during low flow conditions and small runoff duration. Throughout the stormwater event, manual grab sampling was obtained for all monitoring sites considering flow and time. Sampling methodology and interval time for each monitoring site is described below.

3.3.1 Small catchment. It represents the discharge from two or more land use areas. It can be one outflow to a receiving waterbody or a stream channel. In this type of monitoring site, usually two cases need to be considered: i.e. time-spaced sampling (due to surrounding impervious urban areas) and flow weighted sampling (due to discharge from agriculture, construction and other pervious areas). In this study, average lag time between the start of rainfall and start of runoff was early start of runoff due to urban site impact and peak concentrations were observed with increase in flow rate in later period of runoff for all stormwater events in 2012, 2013 as shown in Figure 2 (a) and (b). Due to surrounding urban land use impact the initial peak concentration was obvious whereas later peak concentration due to runoff flow from concentration site. Moreover, it was perceived after few hours of start time that runoff from pervious areas discharge impact occurred and sustained for limited hours consequently sampling was nonstop till the runoff becomes clear and clean and also the flow in runoff is reduced.

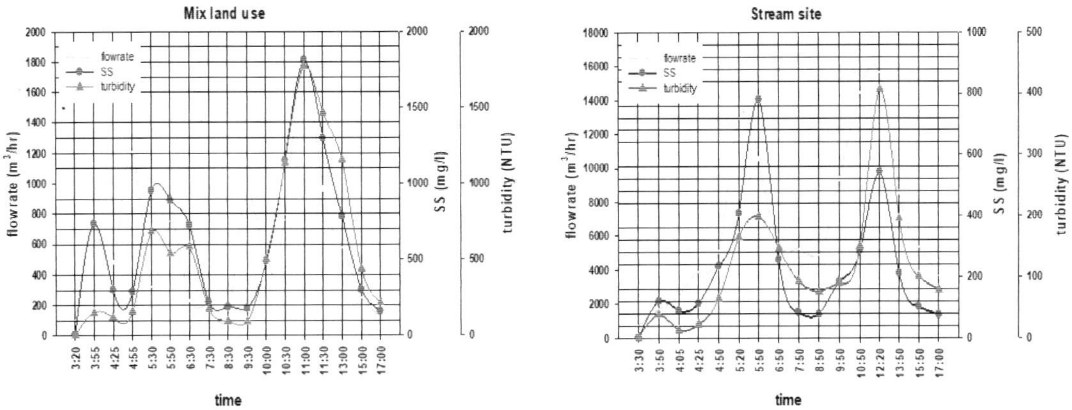

Figure 2. (a) Discharge from mix land use (b) Discharge from stream site

3.3.2 Construction site. Due to devegetated ground land, deposition, infiltration, and soil disturbance activities the average lag time between the start of rainfall and start of runoff was observed more than 3 hours. The peak flow rate was preceded by the peak concentration but in the year 2012 it was not correlated and it was correlated in the year 2013. It occurs due to phase construction and drainage pattern activities changed very much.

The stormwater events with no base flow conditions resulted occurrence of peak pollutants concentration within 2 hours of initial runoff in the year 2012 as shown in pollutograph, therefore time spaced technique in 1st part of runoff can be considered, whereas, results with base flow conditions in 2013 showed peak concentration in the later period of rainfall event and flow spaced technique can be considered in overall event as shown in Figure 3 (a) & (b).

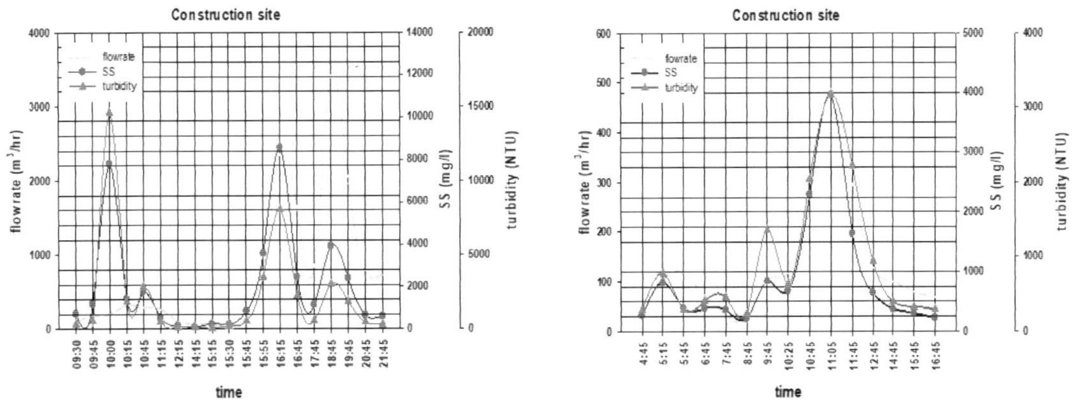

Figure 3. (a) Runoff in 2012 (b) Flow condition in 2013

3.3.3 Urban Site. Urban area delivers the discharge for this site. Average lag time was less than 30 minutes of rainfall between the start of rainfall and start of runoff. Due to high imperviousness and urban land use impact early flushing peak concentration was observed as shown in Figure 4 (a) & (b). So, it is important to capture initial stormwater runoff sampling and hence it is considered by time weighted sampling. Pollutants concentration trend was decreased slowly and consequently quantity of samples can be decreased in middle and later period of storm event.

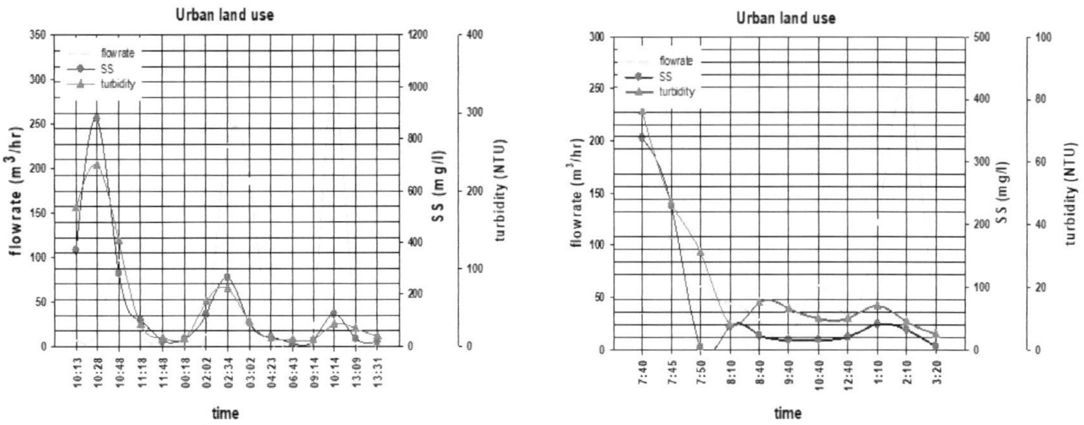

Figure 4. (a) Discharge from urban area 2012 (b) Discharge from urban area 2013

4. Conclusions

The two major runoff effects including stormwater quantity and quality by urban expansion leads in change from natural land and pervious cover to impervious cover and unnatural land. In addition, it increases peak flow and runoff volumes within the watershed; globally it is a common aspect of urban development. This study summarizes the understanding of stormwater monitoring plan, sampling strategies and techniques, determining the trends for problem parameters and evaluating the hydro-pollutograph trend according to site-specific conditions. This research provides better understanding of temporal variability for hydrological and water quality variables during inter-intra storm events.

5. References

[1] Shuster W D, Bonta J, Thurston H, Warnemuende E, Smith D R 2005 Impacts of impervious surface on watershed hydrology A Review *Urban Water J.* 2 263-275.

[2] Hall M J, 1984 Urban Hydrology *Elsevier Applied Science Publishers New York, USA.*

[3] Law N L, McNeal L F, Cappiella K, Pit R, 2008 Monitoring to Demonstrate Environmental Results Guidance to Develop Local Stormwater Monitoring Studies Using Six Example Study Designs.

[4] Carltrans Stormwater Guidance Manual, 2013 Developing the Study Plan California Department of Transportation, Document No. CTSW-OT-13-999.43.01.

[5] Stormwater Discharge Monitoring Plan consists of Sampling and Analysis Plan and Quality Assurance Project Plan 2012 *Water Pollution Control Facilities* City of Portland.

[6] Yoon S W, Chung S W, Oh D G, Lee J W, 2010 Monitoring of non-point source pollutants load from a mixed forest land use *J. Env. Sci.,* 22, 801-805.

[7] Chang M, McBroom M W, Beasley R S, 2004 Roofing as a source of nonpoint water pollution *J. Env. Manag.,*73 307-315.

[8] Vincent S N, Neel T C, and Dewey J B, 2012 Pollutant Washoff and Loading from Parking Lots *Ninth International Conference on Urban Drainage* United States

[9] Priyantha S, 2011 Characteristics of stormwater runoff in different land use area L. (ed) International conference on Environmental science and Development pp 301-301

[10] Maniquiz M C, Lee S Y, Kim L H, 2010 Long-Term Monitoring of Infiltration Trench for Nonpoint Source Pollution Control *Water Air & Soil pollution* 212 13-26.

[11] Grayson R B, Moore I D, McMahon T A, 1992 Physically based hydrologic.

[12] U S EPA, 2003 Developing water quality criteria for suspended and bedded.

[13] Washington, 2004.

[14] Francey M, 2010 Characterizing Urban Pollutant Loads *PhD thesis Department of Civil Engineering,* Monash University, Australia.

IOP Conf. Series: Earth and Environmental Science **127** (2018) 012015 doi:10.1088/1755-1315/127/1/012015

Advanced oxidation of biorefractory organics in aqueous solution together with bioelectricity generation by microbial fuel cells with composite FO/GPEs

Bao-rong FU[1], Chao SHEN[1], Jing REN[1*], Jia-yi CHEN[2], Lin ZHAO[2]

[1]School of Environment Science, Liaoning University, Shenyang 110036, China
[2]Department of Environmental Science and Engineering, Tianjin University, Tianjin 300072, China

Corresponding Author E-mail: renjing@lnu.edu.cn (Ren J.)

Abstract. In this study, ferric oxide loading graphite particle electrodes (FO/GPEs) were prepared as cathode of a three-dimensional electrode MFC-Fenton system. The properties of the composite cathode were examined with higher surface area and more mesopores. FO/GPEs could work as both cathode and Fenton iron reagents, contributing to high oxidation activity and better performance of electricity generation. The application of FO/GPEs MFC-Fenton system on degrading p-nitrophenol presented high catalytic efficiency in a wide range of pH value. The removal of p-nitrophenol and TOC attained to about 85 % within 8 and 64 h at neutral pH, respectively. A neutral FO/GPEs MFC-Fenton oxidation mechanism was also proposed. Specifically, both the surface iron sites and dissolved iron ions catalyzed the decomposition of H_2O_2. As results, the generated hydroxyl radicals were used for p-nitrophenol degradation and the iron oxide was recycled.

1. Introduction

Concerns for environmental impact along with enhanced public awareness of effects of contamination by hazardous chemicals, has led to strict environmental controls. Technological progress and increased utilization of natural resources have resulted in larger and more diverse discharge of organics to the environment. Of these, bio-refractory organic substances are of particular concern as they resist degradation and can accumulate in the environment. Advanced oxidation process (AOP) based on Fenton's reaction as a favorable method has been widely used for degradation of bio-refractory organic with the generation of hydroxyl radicals (•OH) from Fenton's reagents (Fe^{2+} and H_2O_2)[1]. The electro-Fenton process is a recent development which offers many advantages over the chemical Fenton approach because of the higher efficiency of Fenton's reagents utilization and less risk of chemical transport and storage [2]. However, it will have high power consumption leading to higher energy cost [3].

Microbial fuel cells (MFCs) are a newly developed bioelectrochemical system which generates electricity and simultaneously treats wastewater, MFCs presently enlarge the application range to be a device synthesizing H_2O_2 through using bio-electrons to occur two-electron oxygen reduction at carbonaceous cathode but with lower bioelectricity output [4]. Particularly, with the presence of ferrous iron, the *in situ* generated H_2O_2 by MFCs further reacts with ferrous iron to produce hydroxyl radicals, establishing a better energy-generation system with biodegradable organics elimination in anode and biorefactory organics removal in cathode. Scrap iron and natural limonite were widely used as the iron sources to conduct Fenton's reaction in MFCs for organic pollutant removal in former research [3, 5]. The oxidation of pollutants is most efficient at acidic conditions in Fenton system using free irons, however, it will become weak when the pH is over 4 [6]. The formation of iron sludge attributed by pH adjustment limits the wide application and continuous addition of ferrous ion enormously increases the operation cost [2]. In this case, it is meaningful to investigate the Fenton reaction at neutral pH for a wider industry implementation. To overcome these shortages, loading iron oxides on carboneus cathode to fabricate a composite cathode has been used in MFC-Fenton system for water treatment, e.g. CNT/γ-FeOOH composite cathode [2] and Fe@Fe_2O_3/carbon felt composite cathode [1, 7].

It is noticeable that as a H_2O_2-producing device, the MFC with the three-dimensional electrode is more attractive than a two-dimensional one. It is mainly because that particle electrodes form charged microelectrodes which may increase mass transfer coefficient [8] and enhance energy production as well [4]. Graphite particle electrodes (GPEs) described in previous research [4, 9] show a high efficient cathodic catalyst for the *in situ* generation of H_2O_2 due to its large surface area, good conductivity and superior electrochemical activity of the two-electron oxygen reduction. Moreover, due to its excellent adsorption, GPEs can easily adsorb iron ions to fabricate a composite electrode. The ferric oxides (FO) are widely used and functioned as iron sources of the MFC-Fenton reactions [10, 11]. As for composite cathode, FO *in situ* generates adsorbed ferrous ion and then is dissolved to aqueous solution as Fe^{2+}. A constant amount of iron ions can be self supplied by iron oxide in composite cathode and sustainably recycled after treatment.

In this study, the newly composite FO/GPEs were fabricated for the MFC-Fenton system. p-nitrophenol (PNP) was chosen as a representative of bio-refractory organic pollutant. The aim was to demonstrate the feasibility of using such a MFC-Fenton system with composite FO/GPEs as the cathode to degrade p-nitrophenol in aqueous solution under neutral condition.

2. Materials and Methods

2.1. Fabrication of composite FO/GPEs

Composite FO/GPEs in present study were prepared by loading ferric oxides onto the graphite particle electrodes. Firstly, the graphite particle electrodes were made by granulation of graphite powders and PTFE with a diameter of 2-3 mm, according to procedures described in previous studies [4, 12]. Subsequently, $FeCl_3$ solution (solution A; 0.1 mol·L^{-1}) and NaOH solution (solution B; 0.5 mol·L^{-1}) were prepared for loading process [13]. Solution A flew through a beaker packed with the graphite particle electrodes to preload the ferric ions. Then, solution B was used to make ferric iron precipitated onto inner surface of graphite granules. The ferric oxide loaded graphite particle electrodes was washed with deionized water and finally dried in nitrogen to fabricate ferric oxide/graphite particle electrodes (composite FO/GPEs). The quantity of Fe loading on the GPEs was measured by the weight method.

2.2. Characterization and porosity of FO/GPE composite cathode

The surface morphology of carbon powders, graphite particle electrodes and the FO/GPEs were characterized using scanning electron microscope (SEM; Nanosem 430, FEI Electron Optics Company, USA). The Fe elemental analysis of FO/GPEs was carried out by XPS (PHI1600, Physical Electronics, USA) to determine the properties of Fe element. The physical properties of carbon powders, graphite particle electrodes and composite FO/GPEs were measured using multipoint Brunauer-Emmett-Teller (BET) surface area method and Barrett-Joyner-Halenda (BJH) pore size and volume analysis through N2 adsorption/desorption analysis on Autosorb-1 (Quantachrome Instruments, USA).

2.3. MFC configuration and experiment operation

MFC-Fenton degradation of p-nitrophenol was performed in a dual-chamber MFC with composite FO/GPEs. The anodic and cathodic compartments were separated by a cation exchange membrane (Ultrex CMI-7000, Membranes International Inc., Glen Rock, NJ, USA) into the volume of 32.0 mL and 64.0 mL, respectively. The anode was carbon felt (3.0 cm length × 3.0 cm width, Hesen, Shanghai,

China), which was soaked in acetone for 24 h to remove potential organics from surface . FO/GPEs were filled in the cathode chamber with a graphite rod (0.8 cm diameter and 6cm length) as cathode to provide the external contact, reducing cathodic working volume of 50.0 mL. Cathode solution was 0.05 mol·L−1Na_2SO_4 (pH=7.0) fed by fresh air. The anode was set up following our previous work[4, 9]. As a comparison, three-dimensional electrode MFCs with only GPEs and 0.1g·L−1 Fe2+ and GPEs (Fe2+/ GPEs) were conducted under the same conditions. In the stage of start up, an external resistor of 1000 Ω was used to connect the MFC circuit. After inoculation, MFCs ware operated under open-circuit condition at least three successive batch cycles with a stable maximum voltage produced. Subsequently, all MFCs were changed the cathode solution (50 mmol·L−1 Na_2SO_4 solution containing 50 mg·L−1 PNP (pH=7.0)) bubbled with air before degradation experiments. The external resistor set up at 20 Ω. 0.5 mol·L−1 of NaOH or H2SO4 were used to adjust the solution pH. The system was then stirred for 30 min to establish adsorption/desorption equilibrium between the solution and electrodes in the cathode chamber [10]. All experiments were conducted in an incubator at 30±0.5 °C.

2.4. *Analytical methods*
The cell voltages were monitored in one minute interval by a data acquisition system (PISO-813, ICP DAS Co., Ltd, China) and recorded by a personal computer. The polarization and power density curves used to evaluate the power performance of MFCs, which were performed by varying the external resistances over a range of 5-5000 Ω included in a resistor box during hours when stable maximum voltages produced in a steady and repeatable voltage cycle. The anode and cathode potentials were measured by saturated calomel electrodes (SCE, +0.242 V vs. SHE) for reference. The concentrations of *p*-nitrophenol were detectd using high performance liquid chromatography (HPLC, Waters E 2695 Separation Module 2998 PDA, Waters, USA) with a C18 column (5 μm, 4.6×250 mm) and a UV detector. The mobile phase was methanol/ water (60/40) and the flow rate was 1.0 $mL\cdot min^{-1}$. The UV detector was set at 320 nm[14]. The dissolved iron ion concentrations were colorimetric determined using 1,10-phenanthrolin at 510nm [15]. Total organic carbon (TOC) of initial and treated aqueous PNP solutions were carried out by TOC Analyzer (TOC-VCPH, Shimadzu Corporation, Japan) equipped with an automatic sample injector. The concentration of H_2O_2 was measured by spectrophotometry of Ti^{4+}-H_2O_2 complex at 410nm [16].

3. Results and Discussion

3.1. *Characterization of composite FO/GPEs*
The morphological characterization of powders, graphite-PTFE granules (GPEs) and FO/GPEs were scanned by SEM and the results are presented in Fig. 1. Graphite powders had an obvious sheet-like morphology and random pore size distributions (Fig. 1a). The SEM images (Fig. 1b, 1c) show that more channels and pores were generated by bonding graphite powders with massive tight and brawny PTFE fibers in graphite-PTFE granules and FO/GPEs. The surface morphology of FO/GPEs, which was loaded by ferric oxides, was much rougher than graphite-PTFE granules (GPEs).

(a) (b) (c)

Figure. 1 SEM images of (a) graphite powders, (b) GPEs and (c) FO/GPEs at a magnification of 10 K

The XPS analysis verified the coexistence of C, O and Fe in FO/GPE composite cathode, which demonstrated the successful preparation of the composite FO/GPEs with 9.2 % weight of ferric contents. In addition, oxygen content of FO/GPEs (14.2 %) was higher than that of GPEs (7.3 %), which indicated that the ferric element was loaded on composite cathode mainly in the form of ferric oxides. The further analysis of Fe 2p spectra was carried out to determine the properties of Fe element on FO/GPEs. The results demonstrated the existence of FeO (709.3 eV)[16] and α-Fe_2O_3 (711.4 eV) [17] on composite electrode.

3.2. Porosity of composite FO/GPEs
The porous properties and the structure on pore size distribution were measured by BET and BJH technology. N_2 adsorption/desorption isotherms of graphite powders, GPEs and FO/GPEs all belonged to type IV with a pronounced hysteresis at higher P/P_0 [18]. The type H4 hysteresis loops in isotherms indicated the existence of slit-like mesopores [19] in three samples, corresponding to the morphology shown in the SEM images. Detailed information regarding porous properties is summarized in Table 1. As expected, the absorption volume of GPEs and FO/GPEs were relatively high when compared to that of graphite powders due to the larger surface area of GPEs and FO/GPEs than graphite powders. More channels and pores were generated by plenty of binder fibers in graphite-PTFE granules, which led to the larger surface area and more mesopores of GPEs (46.9 $m^2 \cdot g^{-1}$ and 28.2 $m^2 \cdot g^{-1}$) than graphite powders(38.7$m^2 \cdot g^{-1}$ and 21.3 $m^2 \cdot g^{-1}$). Simultaneously, graphite-PTFE granules loaded with ferric oxides (FO/GPEs) had higher surface area (68.6 $m^2 \cdot g^{-1}$) and larger pore volume (0.049 $m^3 \cdot g^{-1}$) than unloaded graphite-PTFE granules (GPEs, 46.9 $m^2 \cdot g^{-1}$ and 0.043 $m^3 \cdot g^{-1}$) due to the ferric oxides dispersed on the inner and external of graphite-PTFE granules. More micropores and mesopores were also formed by loading iron oxides on GPEs, corresponding to the rougher surface of FO/GPEs in SEM images. The results of XPS clearly demonstrate that the higher percentage of mesoposity was presented in FO/GPEs over GPEs, which would strongly increase the metal leaching and further enhance the catalytic performance [20].

Table 1. Porous structural characteristics of GPEs and FO/GPEs

Sample	graphite powders	GPEs	FO/GPEs
$S_{BET}{}^a$ ($m^2 \cdot g^{-1}$)	38.7	46.9	68.6
Microspore area ($m^2 \cdot g^{-1}$)	17.4	18.7	21.7
Mesopore area ($m^2 \cdot g^{-1}$)	21.3	28.2	46.9
Total pore volume($cm^3 \cdot g$-1)	0.041	0.043	0.049
Microspore volume ($cm^3 \cdot g$-1)	0.024	0.020	0.021
Mesopore volume ($cm^3 \cdot g$-1)	0.017	0.023	0.027
Average pore diameter (nm)	4.936	5.233	7.301

[a] Specific surface area

3.3. Simultaneous electricity generation in MFC using composite FO/GPEs as cathode
The power density and polarization curves of MFCs using GPEs, Fe^{2+}/GPEs and FO/GPEs as cathode were tested by varying external resistance from 5 Ω to 5000 Ω and compared in Fig. 2. MFC-FO/GPEs produced the highest open circuit voltages (OCV) of 0.555 V, followed by the MFC-Fe^{2+}/GPEs of 0.527 V and then the MFC-GPEs of 0.477 V (Fig. 2A). No obvious activation polarization loss was displayed in MFCs. The maximum power density of MFC-FO/GPEs (5.11 $W \cdot m^{-3}$) and MFC-Fe^{2+}/GPEs (4.66 $W \cdot m^{-3}$) were 28.0 % and 40.4 % higher than that of MFC-GPEs (3.64 $W \cdot m^{-3}$) (Fig. 2A). The external resistance of all MFCs was 200 Ω. According to the polarization curves (Fig. 2B), the differences of electricity outputs were contributed by the differences in cathode potentials since the anode potentials of all MFCs were similar. An excellent power output was produced by MFC using FO/GPEs which had a higher specific surface. Additional ORR sites were provided by more micropores and mesopores in FO/GPEs, contributing to a better performance than GPEs. Meanwhile, the existence of Fenton iron reagent improved the MFC performance mainly due to the high redox

potential of •OH from Fenton's reaction [21]. The increasing electricity generation of MFC-FO/GPEs over MFC-Fe^{2+}/GPEs was caused by the more kinetic favorability of leached Fe^{2+} from FO/GPEs than Fe^{2+} in the bulk solution in MFC-Fenton system [7].

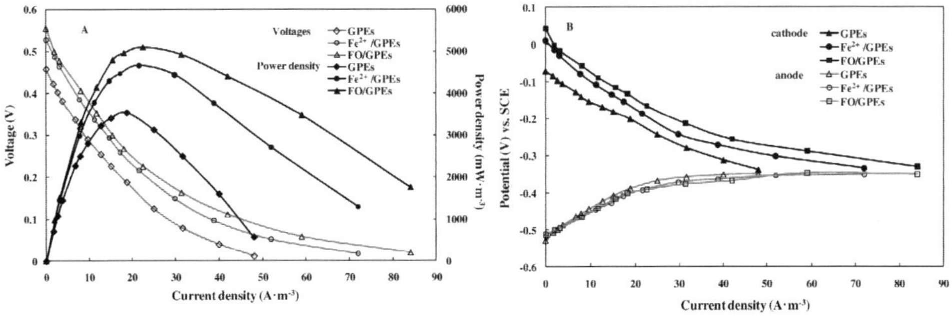

(a) (b)

Figure. 2 (a) Power density (normalized to the cathode effective) and Cell Voltages as a function of the current density in MFCs using GPEs, Fe^{2+}/GPEs and FO/GPEs; (b) Electrode potentials for the MFCs using GPEs, Fe^{2+}/GPEs and FO/GPEs.

3.4. The degradation of p-nitrophenol in FO/GPEs MFC-Fenton system

The degradation of PNP during 8 h in MFC-GPEs process, Fe^{2+}/GPEs MFC-Fenton system and FO/GPEs MFC-Fenton system at pH 3 and pH 7 was displayed in Fig. 3. Without any Fenton iron reagents in MFC, the electrochemical degradation of PNP on GPEs were merely 17.6 % (pH=3) and 14.4 % (pH=7) in 8 hours. The low degradation efficiencies might be attributed to cathode reduction on GPEs and relative weaker oxidation of H_2O_2 [5]. Simultaneously, the FO/GPEs MFC-Fenton system achieved better performance of PNP over Fe^{2+}/GPEs MFC-Fenton system, which indicated that leached Fe^{2+} from FO/GPEs were more favor to involve in Fenton reaction than free Fe^{2+} in solution. The experimental data from Fig. 3 was replotted using a logarithmic linear scale (Fig. 4), kinetic constants and coefficients (R^2) of which were summarized in Table 2. The PNP decomposition followed the pseudo-first-order kinetic reaction. The kinetic constant of FO/GPEs MFC-Fenton system at pH 7 was 0.267 h^{-1}, close to those at acid condition(0.296 h^{-1}), which confirmed the feasibility of FO/GPE MFC-Fenton system at neutral pH. Fig. 5 shows the removal of PNP and TOC in Fe^{2+}/GPEs MFC-Fenton system and FO/GPEs MFC-Fenton system at different pH values. The removal of PNP and TOC quickly declined after 8h and 64 h in Fe^{2+}/GPEs MFC-Fenton system from pH 3 to 7, since the acidity is the optimal reaction condition for hydroxyl radical generation by ferrous iron catalyzing hydrogen peroxide. While both a slight decrease of 2.6 % and 5.1 % in PNP degradation and TOC removal were observed from an acidic FO/GPEs MFC-Fenton system to a neutral process. This is possibly due to the coexistence of the mineral surface-catalyzed heterogeneous reactions on composite electrode to generate oxidants. The removal of PNP and TOC at pH 7 reached 87.9 % in 8 h and 82.1 % in 64 h, respectively, which were even larger than those using Fe^{2+}/GPEs at pH 3, revealing the high activity of FO/GPEs MFC-Fenton system for PNP removal in a wide pH range.

Table 2 Pseudo-first-order kinetic constants and the corresponding coefficients of PNP degradation

pH	Reactor	Kinetic constant(h^{-1})	Coefficient(R^2)
3	MFC-GPEs	0.020	0.981
3	MFC- Fe^{2+}/GPEs	0.201	0.996
3	MFC -FO/GPE	0.267	0.995
7	MFC-GPEs	0.025	0.975
7	MFC- Fe^{2+}/GPEs	0.121	0.984
7	MFC -FO/GPE	0.296	0.993

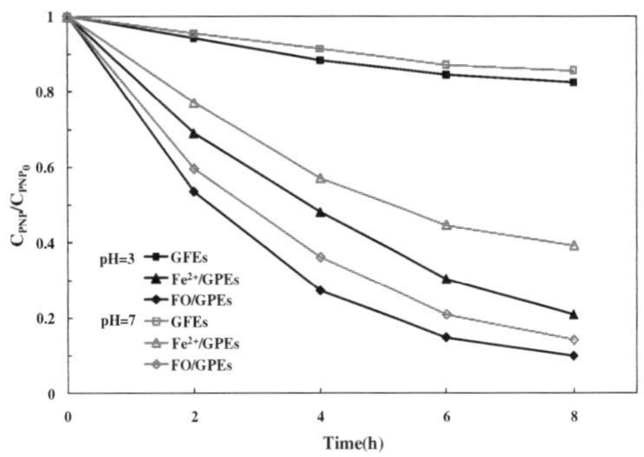

Figure. 3 PNP degradation in MFC-GPEs, Fe^{2+}/GPEs MFC-Fenton system and FO/GPEs MFC-Fenton system at neutral pH (pH=7) and acidic conditions (pH=3).

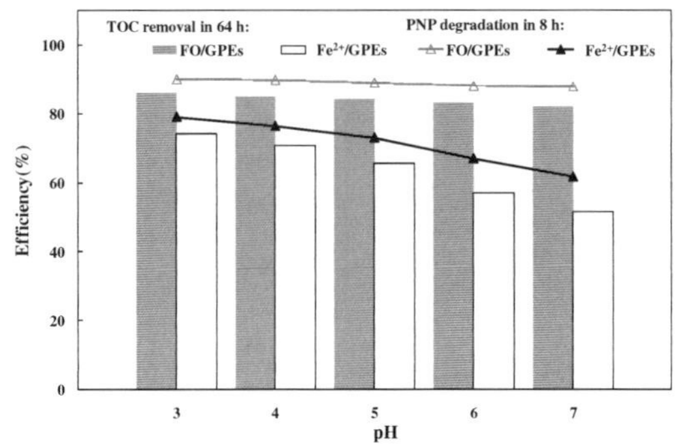

Figure. 4 TOC removal (in 64 h) and PNP degradation (in 8 h) in Fe^{2+}/GPEs MFC-Fenton system and FO/GPEs MFC-Fenton system at different pH from 3 to 7.

To understand the high degradation efficiency of this MFC-Fenton process at neutral pH, the concentrations of hydrogen peroxide and iron ions produced during PNP degradation at pH 3 and pH 7 were measured and the results are displayed in Fig. 5. The H_2O_2 concentration started to increase with the reaction time and then stabilized after approximately 32 hours due to the accumulation. The decomposition and generation processes of hydrogen peroxide reached a balance in the cathode chamber during the last half period. The production of H_2O_2 at acidic condition was much less than that at neutral pH for the reason that the optimum pH value for the electrochemical generation of hydrogen peroxide was around 10[22]. Moreover, the concentration of iron ions after 64 h in the FO/GPEs MFC-Fenton process at pH 7 (12.5 mg·L^{-1}) was 74.7 % lower than that at pH 3 (about 49.5 mg·L^{-1}). This is mostly because that the leaching process of iron ion from FO/GPEs to solution prefers to occur under acidic condition (Eq. (1)). The concentration of iron ions in FO/GPEs MFC-Fenton system under neutral condition only increased quickly at the beginning but did not rise any more after 24 h, demonstrating that H_2O_2 decomposition at higher pH trended to a surface-catalyzed process. This result suggested that the composite FO/GPEs could be reused.

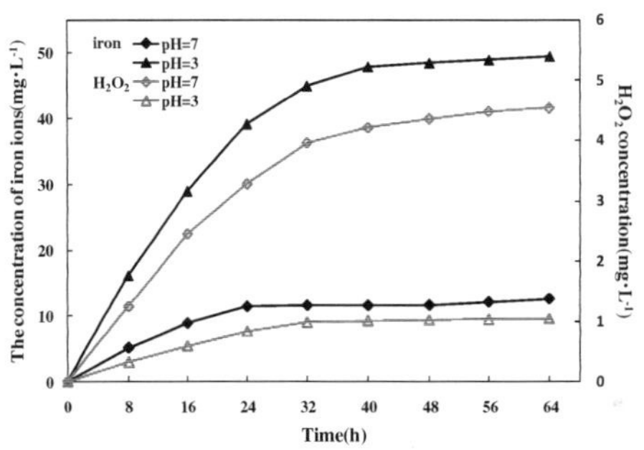

Figure. 5 The concentrations of hydrogen peroxide and iron ions produced during PNP degradation in FO/GPEs MFC-Fenton system at pH 3 and pH 7.

3.5. The proposed mechanism of FO/GPEs MFC-Fenton system

On the basis of the results above, a possible mechanism for the PNP degradation in the FO/GPEs MFC-Fenton system was proposed. Specifically, acetate was oxidized by electricigens in anode chamber to release electrons and protons. After that, the electrons collected in the anode were conducted through the external resistor to the cathode and the cation migrated through the proton exchange membrane into the cathode chamber. In cathode chamber, the dissolved oxygen was firstly adsorbed on the surface of FO/GPEs and immediately electrochemically reduced to produce H_2O_2. At the same time, ≡Fe(II) and ≡Fe(III) species were corroded from FO/GPEs, which could be depicted by the reaction where ≡Fe(II) and ≡Fe(III) species stands for Fe(II) and Fe(III) sites on the catalyst FO surface. The initially generated ≡Fe(II) further reacted with H_2O_2 to generate surface-bound $\bullet OH_{ads}$ on FO/GPEs [23]. Simultaneously, the formed ≡Fe(II) and ≡Fe(III) on FO/GPEs was leached into the cathodic solution, but the porous structure in FO/GPEs might absorb partial Fe^{2+} and Fe^{3+} to prevent the complete leaching [2, 7]. The absorbed iron ($Fe^{2+}{}_{ads}$) could catalyze the decomposition of H_2O_2 both on FO/GPEs and in bulk solution to generate $\bullet OH_{ads}$ and $\bullet OH_{aq}$, respectively. Also, the dissolved iron ($Fe^{2+}{}_{aq}$) was dispersed into bulk solution to initiate the decomposition of H_2O_2 to $\bullet OH_{aq}$. The $\bullet OH_{ads}$ from the surface of FO/GPEs might be minimally diffused into the bulk. Finally, PNP was degraded mainly by hydroxyl radicals including •OH on the surface of FO/GPEs and in bulk solution[3, 14]. The main recycle of ≡Fe(II) from ≡Fe(III) was a direct cathode reaction [11].

Therefore, a reasonable oxidation mechanism of this novel FO/GPEs MFC-Fenton system combined a typical Haber–Weiss mechanism and heterogeneous surface-catalyze reactions was proposed, contributing to the good performance on PNP degradation in this system at neutral pH.

4. Conclusions

A three-dimensional electrode MFC-Fenton system based on the novel composite FO/GPEs as cathode for PNP degradation was investigated in this paper. The prepared FO/GPEs in this study showed multiplicated porosities including micropores and mesopores. The hematite and magnetite were dispersed on the surface of composite FO/GPEs electrode, as indicated by XPS. Higher surface area and more mesoporsity in FO/GPEs increased the efficiency of *in-situ* electro-generation of H_2O_2 and improved Fenton catalytic performance as well. When at the optimum pH 3, FO/GPEs MFC-Fenton system presented higher PNP degradation efficiency compared to the system with Fe^{2+} as the iron reagents. As pH increased to neutral, this system kept in good performance without sacrificing efficiency. The removal of PNP and TOC reached respectively around 85 % within 8 h and 64 h, respectively. Finally, the oxidation mechanism of MFC-Fenton system with novel FO/GPEs combined a typical Haber–Weiss mechanism and heterogeneous surface-catalyze reactions was proposed and showed good performance on the PNP degradation in this system at neutral pH.

5. References

[1] Zhuang L, Zhou S, Yuan Y, et al. (2010) A novel bioelectro-Fenton system for coupling anodic COD removal with cathodic dye degradation *Chemical Engineering Journal*, 163, pp.160-163.
[2] Feng C-H, Li F-B, Mai H-J, et al. (2010) Bio-Electro-Fenton Process Driven by Microbial Fuel Cell for Wastewater Treatment. *Environmental science & technology*, 2010, 44, pp.1875-1880.
[3] Tao H-C, Wei X-Y, Zhang L-J, et al. (2013) Degradation of p-nitrophenol in a BES-Fenton system based on limonite. *Journal of hazardous materials*, 254-255, pp.236-241.
[4] Chen J-y, Li N, Zhao L. (2014) Three-dimensional electrode microbial fuel cell for hydrogen peroxide synthesis coupled to wastewater treatment. *Journal of Power Sources*, 254, pp.316-322.
[5] Zhu X, Ni J. (2009) Simultaneous processes of electricity generation and p-nitrophenol degradation in a microbial fuel cell. *Electrochemistry Communications*, 11, pp.274-277.
[6] Smith B A, Teel A L, Watts R J. (2004) Identification of the reactive oxygen species responsible for carbon tetrachloride degradation in modified Fenton's systems. *Environmental science & technology*, 38, pp.5465-5469.
[7] Zhuang L, Zhou S, Li Y, et al. (2010) In situ Fenton-enhanced cathodic reaction for sustainable increased electricity generation in microbial fuel cells. *Journal of Power Sources*, 195,pp. 1379-1382.
[8] Wei L, Guo S, Yan G, et al. (2010) Electrochemical pretreatment of heavy oil refinery wastewater using a three-dimensional electrode reactor. *Electrochimica Acta*, 55,pp.8615-8620.
[9] Chen J-y, Zhao L, Li N, et al. (2015) A microbial fuel cell with the three-dimensional electrode applied an external voltage for synthesis of hydrogen peroxide from organic matter. *Journal of Power Sources*, 287,291-296.
[10] Li J, Ai Z, Zhang L. (2009) Design of a neutral electro-Fenton system with $Fe@Fe_2O_3$/ACF composite cathode for wastewater treatment. *Journal of hazardous materials*, 164, pp.18-25.

[11] Zhao H, Wang Y, Wang Y, et al. (2012) Electro-Fenton oxidation of pesticides with a novel Fe_3O_4@Fe_2O_3/activated carbon aerogel cathode: High activity, wide pH range and catalytic mechanism. *Applied Catalysis B: Environmental*, 125, 120-127.

[12] Dong H, Yu H, Wang X. (2012) Catalysis kinetics and porous analysis of rolling activated carbon-PTFE air-cathode in microbial fuel cells. *Environmental science & technology*, 46, pp.13009-15.

[13] Ren J, Li N, Li L, et al. (2015) Granulation and ferric oxides loading enable biochar derived from cotton stalk to remove phosphate from water. *Bioresource technology*, 178, pp.119-25.

[14] Sun S-P, Lemley A T. (2011) p-Nitrophenol degradation by a heterogeneous Fenton-like reaction on nano-magnetite: Process optimization, kinetics, and degradation pathways. *Journal of Molecular Catalysis A: Chemical*, 349, pp. 71-79.

[15] Federation W E, Association A P H. (2005) Standard methods for the examination of water and wastewater. *American Public Health Association (APHA): Washington, DC, USA*, 2005.

[16] Lovley D R, Phillips E J. (1988) Novel mode of microbial energy metabolism: organic carbon oxidation coupled to dissimilatory reduction of iron or manganese. *Applied and environmental microbiology*, 54, pp.1472-80.

[17] Grosvenor A P, Kobe B A, Biesinger M C, et al. (2004) Investigation of multiplet splitting of Fe 2p XPS spectra and bonding in iron compounds. *Surface and Interface Analysis*,36, pp. 564-1574.

[18] Pierotti R, Rouquerol J. (1985) Reporting physisorption data for gas/solid systems with special reference to the determination of surface area and porosity. *Pure Appl Chem*, 57, pp.603-619.

[19] Kumagai S, Ishizawa H, Aoki Y, et al. (2010) Molded micro- and mesoporous carbon/silica composite from rice husk and beet sugar. *Chemical Engineering Journal*, 56, pp.270-277.

[20] Duarte F, Maldonado-Hódar F, Pérez-Cadenas A, et al. (2009) Fenton-like degradation of azo-dye Orange II catalyzed by transition metals on carbon aerogels. *Applied Catalysis B: Environmental*, 85, pp.139-147.

[21] Wood P M. The potential diagram for oxygen at pH 7 (1988). *Biochemical Journal*, 253, pp. 287-289.

[22] Hanaoka K, Sun D, Lawrence R, et al. (2004) The mechanism of the enhanced antioxidant effects against superoxide anion radicals of reduced water produced by electrolysis. *Biophysical Chemistry*, 107, pp.71-82.

[23] Xu L, Wang J. (2012) Magnetic nanoscaled Fe_3O_4/CeO_2 composite as an efficient Fenton-like heterogeneous catalyst for degradation of 4-chlorophenol. *Environmental science & technology,* 46, pp.10145-53.

IOP Conf. Series: Earth and Environmental Science **127** (2018) 012016 doi:10.1088/1755-1315/127/1/012016

An Energy Saving Green Plug Device for Nonlinear Loads

Albe Bloul*, Adel Sharaf, Mohamed El-Hawary

Department of Electrical and Computer Engineering, Dalhousie University, Canada
SHARAF Energy Systems, Inc., Canada

al952598@dal.ca

Abstract. The paper presents a low cost a FACTS Based flexible fuzzy logic based modulated/switched tuned arm filter and Green Plug compensation (SFC-GP) scheme for single-phase nonlinear loads ensuring both voltage stabilization and efficient energy utilization. The new Green Plug-Switched filter compensator SFC modulated LC-Filter PWM Switched Capacitive Compensation Devices is controlled using a fuzzy logic regulator to enhance power quality, improve power factor at the source and reduce switching transients and inrush current conditions as well harmonic contents in source current. The FACTS based SFC-GP Device is a member of family of Green Plug /Filters/Compensation Schemes used for efficient energy utilization, power quality enhancement and voltage/inrush current/soft starting control using a dynamic error driven fuzzy logic controller (FLC). The device with fuzzy logic controller is validated using the Matlab / Simulink Software Environment for enhanced power quality (PQ), improved power factor and reduced inrush currents. This is achieved using modulated PWM Switching of the Filter-Capacitive compensation scheme to cope with dynamic type nonlinear and inrush cyclical loads..

1. Introduction

Fuzzy Logic Control is utilized as the PWM dynamic switching/regulation scheme for the SFC-GP FACTS Device used to enhance power quality and energy utilization for single phase nonlinear loads. It has key advantages compared to other classical PID controller. The advantages of FLC are simplicity, low cost and flexibility in case of uncertain load with unknown mathematical model or process dynamics [1], [2], [3], [4]. Power quality (PQ) problem is defined as any variation in frequency, current or voltage that may lead to an equipment malfunction or failure [5]. In a modern electrical distribution/utilization system, there has been a continuous increase of nonlinear loads, such as uninterruptible power supplies, rectifier equipment used in domestic appliances, communication networks, motors, adjustable speed drives, etc. [6], [7], [8], [9]. Power Quality related problems are of most concern nowadays. The widespread use of electronic equipment, such as programmable logic controllers (PLC), information technology equipment, and power electronics such as adjustable speed drives (ASD), energy-efficient LED-lighting, led to severe nonlinearity and varying range of electric loads nature. All the victims of power quality will be causes by this load. Due to their non-linearity, all these loads cause disturbances in the voltage and current waveform [1]. Whatever the power factor is, however, the generating authority must install machines capable of delivering a particular voltage and current even though, in a specified case, will not most of the voltage and current products is being put to good use also for improving the efficiency, reactive compensation using the capacitors is utilized to improve power factor and reduce harmonic content in Nonlinear- load current [9]. A modulated/switched tuned arm filter is utilized to reduce harmonic currents to reduce cost and simplify

Filter topology [6], [7]. The major thought of filters is to reduce the amount of harmonic in order to improve power quality. The characteristic of passive filters are mainly considered as active alternatives to reduce current harmonic to enhance power factor correction. The disadvantage of passive filters is not modified to follow the dynamic behavior changes of non-linear loads [10]. The advantages of passive filters are terminal voltage regulations, voltage variation reduction, and voltage balance improvement in three phases [11]. The limitations of fixed structure/fixed parameter and detuned passive filters causing series/parallel resonance between system, capacitors and existing AC supply impedance is fully documented. The low-cost passive filter can be utilized with simple switching devices to enhance both dynamics, bandwidth and operation with slow dynamic cyclical and inrush type nonlinear loads. Effective control using PWM- switched power LC-filters can improve effectiveness of passive filter compensation [8], [10]. Enhancement of power quality and improvement of power factor have been proposed in many configurations and literature of filters [12], [2], [3]. Fuzzy system control (FLC) is most suited for control of the pulse width modulation to overcome the negative effects of system nonlinearities and dynamic load variations and variable-uncertainties [13]. A low-cost GP-SFC, C-type scheme has implemented and extended to improve the power quality and efficient utilization in smart utilization grid application. The proposed MNFC- based single phase filter utilizes multi loop dynamic error driven fuzzy logic-FLC to control the SFC-GP FACTS device. The proposed scheme has implemented and succeeded in power factor enhancement and power quality improvement, and full limiting transient over voltage and inrush current conditions. The scheme can be extended to three phase systems [14].

2. Overall system design

A single-phase AC utilization system which is driving an arc type load is shown in figure 1. By analyzing figure 1, we can divide the AC single-phase system into three components namely single-phase power source, the modulated green plug filter and the nonlinear load. The load in the system is the source of nonlinearity and generates harmonics in the power line. Such transients and input current excursions generated by the load need to be compensated and in this study, we have designed a green plug filter to perform this compensation task. The proposed green plug filter is placed in parallel with the load as shown in figure 1. The inner construction of the green plug filter is shown in figure 2. In order to optimally control the operation of the filter, we have deployed a two-input fuzzy logic controller which is inherently a nonlinear controller and is robust to disturbances. The block diagram of the fuzzy logic controller is shown in figure 3 with its inner construction depicted in figure 4. The fuzzy logic controlled green plug filter will make sure that the energy is utilized efficiently, and the power quality is also improved [15].

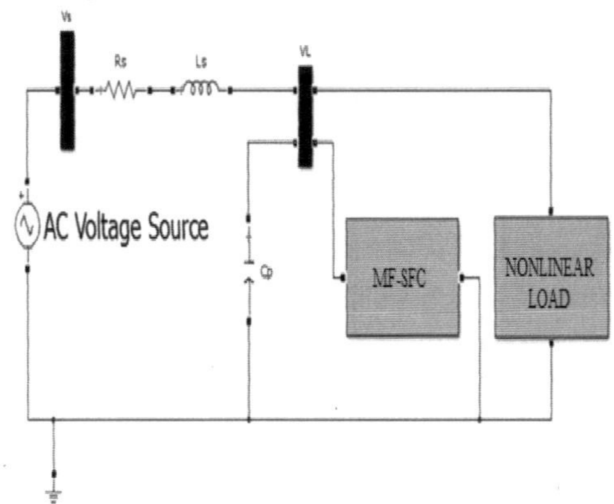

Figure.1. The components of a Single-phase Utility System.

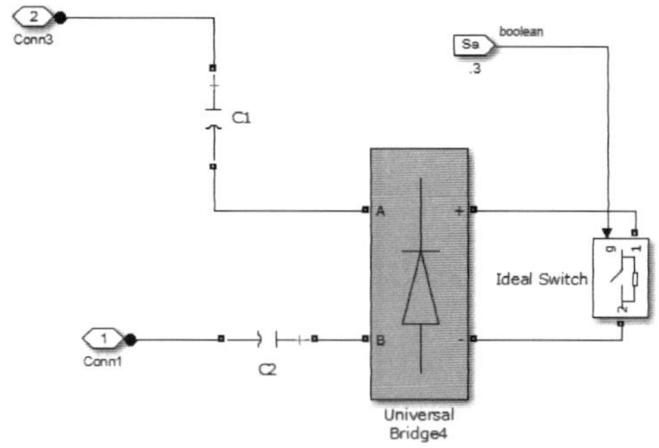

Figure 2. The structure of green plug switched filter

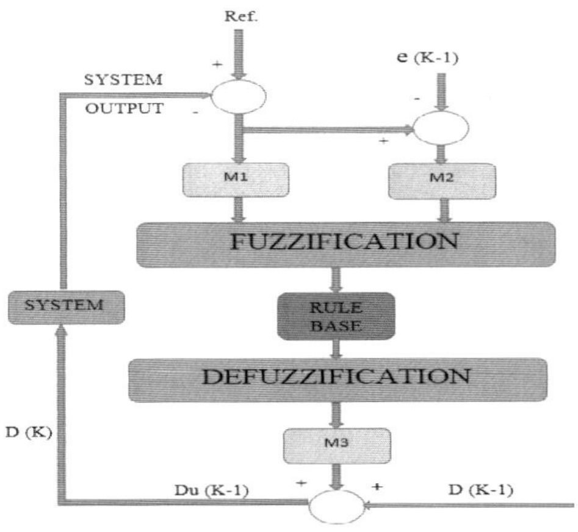

Figure 3. Proposed fuzzy logic controller for green plug filter

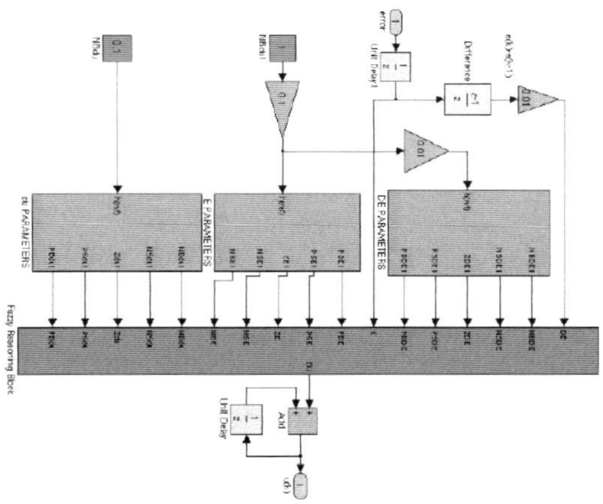

Figure 4. Simulink diagram of fuzzy logic controller

		Δe				
		NB	NS	Z	PS	P
e	PB	Z	PS	PS	PB	PB
	PS	NS	Z	PS	PS	PB
	Z	NS	NS	Z	PS	PS
	NS	NB	NB	NS	Z	PS
	NB	NB	NS	NS	NS	Z

Figure.5. Rule- base design of the fuzzy logic controller

3. Operation of fuzzy logic controller

We will now briefly describe the operation of the two-input, single-output fuzzy logic controller. The fuzzy logic controller accepts the error and change in error as its inputs and produces the control signal for the green plug filter as its output. First, the two inputs are fuzzified using five triangular membership functions. Here, we have preferred triangular membership functions over other types of membership functions such as Gaussian and bell curves as we want to reduce the computational complexity of the controller-driven-filter. After the inputs are fuzzified, a rule base is created to empower the controller with human intelligence. This rule base is shown in figure 5. As can be seen from this figure that all possible combinations of the input membership functions are utilized to construct the rule base. The operation of the rule base can be explained in terms of the sliding mode control theory [16]. As an example, the diagonal line of the rule base is a sliding surface which is the resting state of the system. This sliding surface will be reached after the fuzzy controller will drive the states from other regions towards this surface which is the reaching phase of the system. Once the system will reach the diagonal line, fuzzy logic controller will keep the system onto this line. This means that the equilibrium will be reached and maintained by the fuzzy logic controller. The final stage in fuzzy controller design is known as defuzzification which transforms the linguistic output into a crisp output which can then be fed to the green plug filter. In order to generate the crisp output, we have used Takagi-Sugeno inference mechanism with singleton output variables. The other mechanism is known as Mamdani but it is more complex than the Takagi-Sugeno so we have preferred Takagi-Sugeno keeping in view the online operation of the green plug filter. Here, it is pertinent to mention that the output from fuzzy logic controller are added to its previous output which employs that we are forcing an integral action in the system. The reason of using integral action to eliminate the error completely during the steady state operation of the filter-driven-load system [17].

4. Simulink construction of fuzzy logic controller

We have simulated the complete system in Simulink environment. Here we will briefly discuss the construction of Simulink based fuzzy logic controller. The inputs to the controller are error and change in error. For the change in error, we employ unit delay block and by subtracting the output of the unit delay block from the error signal, we obtain the change in error input signal. Then from the Fuzzy Logic Toolbox in Simulink, we construct the rule base using membership function blocks along with the maximum and minimum operators. Then the output singletons are cut at the minimum level and summed up to generate the incremental control signal. This incremental control signal is also delayed with the help of unit delay block and added with the current control signal to produce the final output. Three scaling elements are also used in the design of fuzzy logic controller. Two of these gain elements are used with the input and one is used with the output. This is to adjust the response of the fuzzy logic controller in relation to the peak values of the error, change in error and the output control signal [18]. Without these gain adjustments, the optimal operation of the filter was not achieved. In this study, we are tuning these factors manually, but this process can be automated and will be considered as a future work.

5. Simulation Results

In order to check the effectiveness of the proposed green plug filter, we conduct the simulations in MATLAB/Simulink environment with several types of the nonlinear loads. The results of the simulation are shown in Figs. 6 through 13. The simulation results include the waveforms for voltage, current, active power, reactive power and the frequency spectrum for each type of load. It can be seen that the green plug switched filter is effective in improving the power quality and reducing the inrush currents due to several types of loads present in the industrial environment.

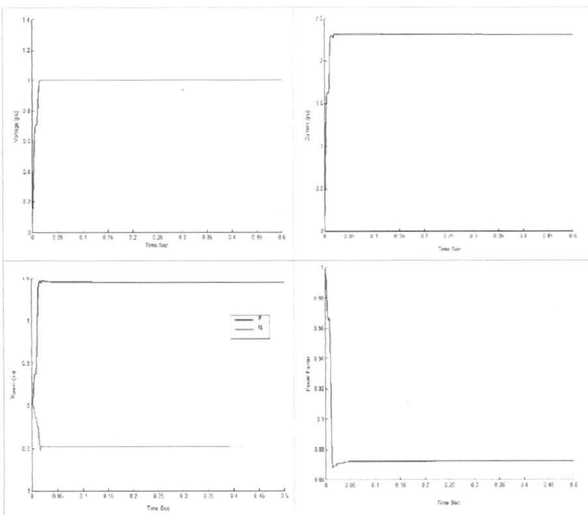

Figure 6. Nonlinear load-RMS V, I, P, Q and PF values at AC source bus.

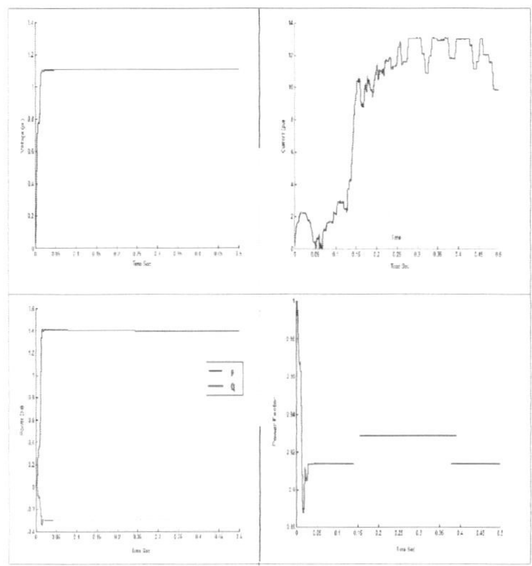

Figure 7. Nonlinear load-RMS V, I, P, Q and PF values at AC load bus

IOP Conf. Series: Earth and Environmental Science **127** (2018) 012016 doi:10.1088/1755-1315/127/1/012016

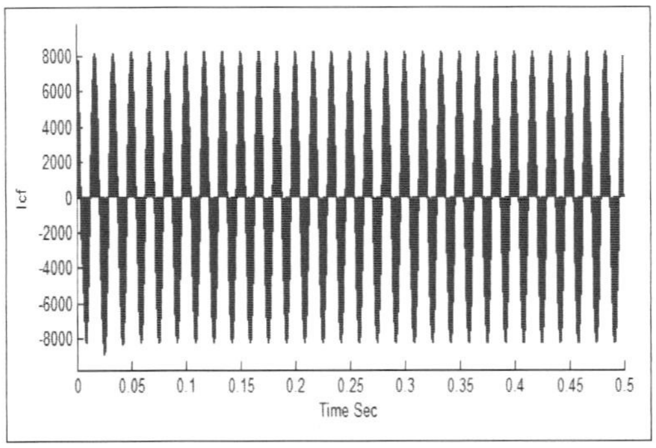

Figure 8. Nonlinear load-Capacitor filter current variation.

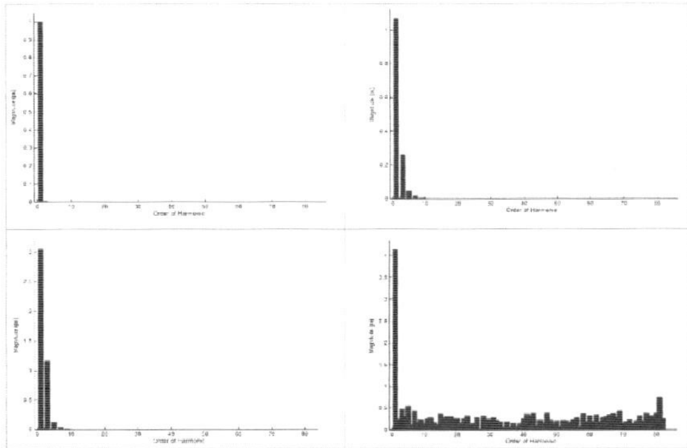

Figure 9. Nonlinear load- FFT frequency spectra of V and I waveforms at both source and load sides

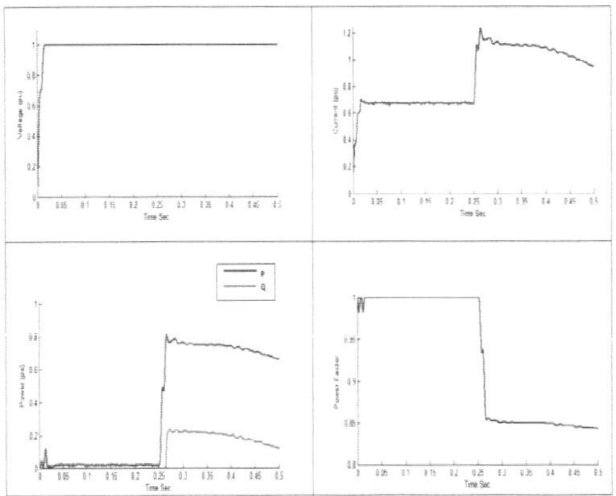

Figure 10. Induction motor cyclic load (full load)- RMS V, I, P, Q and PF values at AC source bus.

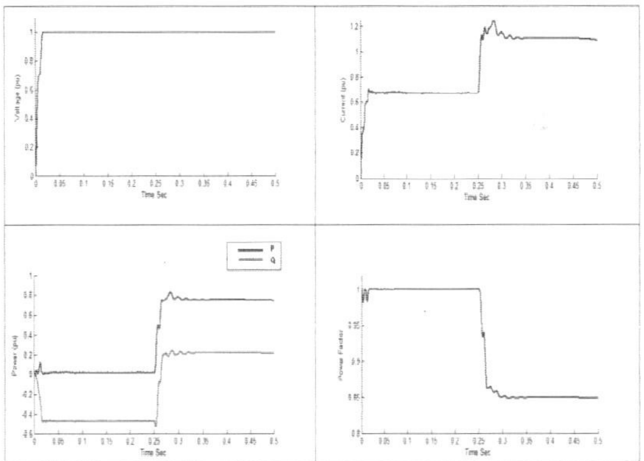

Figure 11. Induction motor cyclic load (full load)- RMS V, I, P, Q and PF values at AC load bus.

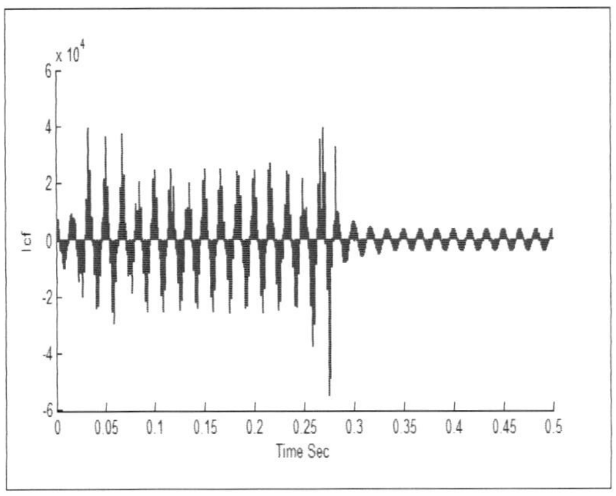

Figure 12. Induction motor cyclic load (full load) Capacitor filter current variation.

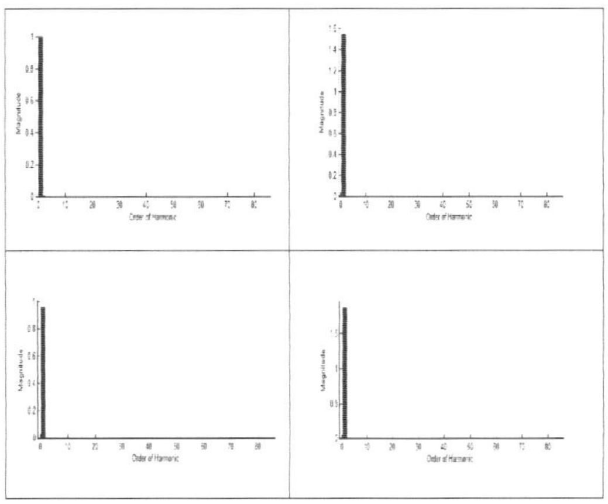

Figure 13. Induction motor cyclic load (full load)- FT frequency spectra of V and I waveforms at both source and load sides

6. Conclusions

This paper presents a FACTS based Green Plug-Switched Filter Compensation SFC_GP scheme controlled by a dynamic multi-loop error driven Fuzzy Logic Controller. The SFC-GP is validated using the Matlab /Simuink Software Environment and the fast-acting Dynamic Fuzzy Logic Controller. The SFC-GP device is controlled by a dynamic tri-loop error driven FLC controller. The SFC-GP Device with Fuzzy Logic Controller is scheme validated for enhanced power quality, voltage stabilization, and efficient utilization. The SFC-GP FACTS Device using a PWM-Switched Filter Compensator scheme is very suitable for nonlinear inrush, dynamic, cyclical and motorized loads efficient electric utilization and power quality/power factor/harmonic reduction. Starting Scheme for voltage stabilization/reduced inrush currents and power factor It can be extended to 3 Phase Nonlinear Loads and utilized for Voltage Dynamic Stabilization with hybrid Smart Grid-Renewable Energy interface schemes using PV/Wind/Fuel to reduce transient voltages, ensure soft starting, limit inrush currents and flicker control. The SFC-GP Device topology can be modified as well as the control strategies to suit specific applications and control duties using fixed topologies as well as modified optimized structures of soft-computing ANN/NFIS/GA/PSO regulators with optimal Gain Scheduling and modified weightings.

7. References

[1] Sharaf, H. Huang and L. Chang 1995 Power quality and nonlinear load voltage stabilization using error driven switched passive power filter *Proc. IEEE International Symposium on Industrial Electronics* pp 616-621

[2] Walid G. Morsi, and M. E. El-Hawary 2008 New Fuzzy-Based Representative Quality Power Factor for Nonsinusoidal Situations IEEE Transcations on Power Delivery **B** 23 L 2.

[3] A. Nur Azlina 2008 Fuzzy Logic Controller for Controlling DC Motor Speed using MATLAB Applications *Power System*.

[4] A. Sharaf and G. Wang 2004 Wind energy system voltage and energy enhancement using low cost dynamic capacitor compensation scheme *Proc. IEEE International Conference on Electrical, lectronic and Computer Engineering* pp. 804-807

[5] I. Altas and A. Sharaf 2007 A generalized direct approach for designing fuzzy logic controllers in matlab/Simulink gui environment *International Journal of Information Technology and Intelligent Computing*.

[6] M. Aredes, K. Heumann, and E. Watanabe 1998 An universal active power line conditioner *IEEE Transactions on Power Delivery* **B** 13 L 2 545-551.

[7] A. Sharaf and R. Chetri 2006 A novel dynamic capacitor compensator/green plug scheme for 3phase-4 wire utilization loads *Proc. IEEE Canadian Conference on Electrical and Computer Engineering* pp. 454-459

[8] M. Rastogi, N. Mohan, and A. Edris 1995 Hybrid-active filtering of harmonic currents in power systems *IEEE Transactions on Power Delivery* **B** 10 L4 1994-2000.

[9] Sharaf, C. Guo, and H. Huang 1995 Distribution/utilization system voltage stabilization and power uality enhancement using intelligent smart filter *Proc. UPEC95*.

[10] P. Ferracci 2000 Power Quality *Schneider Electric Cahier Technique* **B** 199.

[11] I. Vervenne, K. Van Reuse, and R. Belmans 2007 Electric arc furnace modelling from a power quality point of view *Proc IEEE International Conference on Electrical Power Quality and Utilisation* pp. 1- 6

[12] D. Sabinand and A. Sundaram 1996 Quality enhances *IEEE Spectrum* **B** 33 L2 34-41.

[13] H.Fujitaand H.Akagi 1991 A practical approach to harmonic compensation in power systemsseries connection of passive and active filters *IEEE Transactions on Industry Applications* **B** 27 L 6 1020- 025.

[14] Abdelsalam A. Eajal and M. E. El-Hawary 2010 Optimal Capacitor Placement and Sizing in Unbalanced Distribution Systems With Harmonics Consideration Using Particle Swarm Optimization *IEEE Transactions On Power Delivery* **B** 25 L3.

[15] E. Elbakush, A. M. Sharaf, and R. Cakmak 2014 A Novel Fuzzy Logic Control Scheme for FACS-Based Switched Filter Compensation *International Journal of Engineering Innovation & Research* **B** 3 L 2.
[16] H. Fujita, T.Yamasaki, and H. Akagi 2000 A hybrid active filter for damping of harmonic resonance in industrial power systems *IEEE Transactions on Power Delivery* **B** 15 L 2 215-222.
[17] Abdelsalam A. Eajal and M. E. El-Hawary 2010 Optimal Capacitor Placement and Sizing in Unbalanced Distribution Systems With Harmonics Consideration Using Particle Swarm Optimization *IEEE Transactions On Power Delivery* **B** 25 L3.
[18] A. Sharaf 2006 A facts based dynamic capacitor scheme for voltage compensation and power quality enhancement *Proc. IEEE International Symposium on Industrial Electronics* pp. 1200-1205

IOP Conf. Series: Earth and Environmental Science **127** (2018) 012017 doi:10.1088/1755-1315/127/1/012017

Credit Risk Evaluation of Large Power Consumers Considering Power Market Transaction

Li Fulin[2], Xu Erfeng[3], Sun ke[1], Liu Dunnan[3] and Shen Shuyi[4]

[1] State Grid Zhejiang Electricity Power Company, 8 Huanglong Road, Hangzhou, China

[2] Zhejiang Huayun Electric Power Engineering Design Consulting Company, 1 Huadian Road, Hangzhou, China

[3] North China Electric Power University, 2 Beinong Road, Beijing, China

[4] State Grid Zhejiang Economic Research Institute, 1 Nanfu Road, Hangzhou, China

Corresponding Author's E-Mail Address: 18810905481@163.com

Abstract. Large power users will participate in power market in various forms after power system reform. Meanwhile, great importance has always attached to the construction of the credit system in power industry. Due to the difference between the awareness of performance and the ability to perform, credit risk of power customer will emerge accordingly. Therefore, it is critical to evaluate credit risk of large power customers in the new situation of power market. Firstly, this paper constructs index system of credit risk of large power customers, and establishes evaluation model of interval number and AHP-entropy weight method.

1. Introduction

Power industry has always attached great importance to construction of credit system. With the development of power system reform, the types of power market transactions will become more diversified while power market environment will more variable and complex. Therefore, institution of power market regulation should evaluate credit risk of all market subjects urgently, which will provide decision-making basis for effective prevention of market risk and technical support for protection of market standardized operation and healthy and orderly development.

There are many researches on credit index system of power industry. Literature [1] has proposed a customer credit evaluation index system, including the status of industry, the financial situation of enterprises, payment status and so on. Literatures [2-4] have calculated the correlation coefficient between different credit categories. Literature [5] has evaluated the credit risk of power customers based on AHP. In literatures [6-7], power customer risk evaluation model has been constructed by entropy weight method and improved entropy weight method.

In this paper, firstly index system of credit risk evaluation is constructed. Then, based on interval numbers and comprehensive weights, a credit evaluation model is set up, which provides a theoretical reference for the evaluation of large power customer credit risk.

2. Construction of Credit Risk Index System for Large Power Customers

In this paper, credit analysis of large power customers is divided into four parts: external environment, production and operation, financial analysis and transaction credit analysis. As **Table 1** shown, credit evaluation index system for large power customers consists of 4 criterion layers, 9 primary factor layers, and 23 secondary factor layers.

IOP Conf. Series: Earth and Environmental Science **127** (2018) 012017 doi:10.1088/1755-1315/127/1/012017

Table 1. Credit Risk Evaluation Index System for Large Power Customers.

Target Layer	Criterion Layer	Primary Factor Layer	Secondary Factor Layer
Credit Risk of Large Power Customers	Credit Risk of External Environment	Policies and Regulations	Government Support
			Industrial Macro-Control Policies
		Industry Operation Status	Fluctuation of Supply and Demand
			Consolidated Price Increase
	Credit Risk of Production and Operation	Operation Risk	Market Share
		Production Risk	Corporate Social Impact
			Product Qualification Rate
			Safety Accident Rate
	Credit Risk of Financial	Profitability	Return on Assets
			Net Profit Margin on Sales
			Cost Profit Margin
			Asset Liability Ratio
		Solvency	Current Ratio
		Development Ability	Main Business Revenue Growth Rate
			Net Profit Growth Rate
	Credit Risk of Transaction	Direct Purchase of Large Power Consumers	Full Payment Rate
			Time Contribution Ratio
			Contracting Rate of Electricity Sales Contract
			Actual Power Completion Rate
			Contract Performance Rate
		Demand Side Response	Interruptible Load Contract Performance Rate
			Load Control Coordination
			Electricity Check Fit

2.1. External Environment Risk

Enterprises are affected by national environment at a macro level. Besides, the operation of industry and the state related to this industry should also be considered. Therefore, this paper will analysis the project from the aspects of policies and the running state of the industry.

2.2. Production and Operation Risk

The overall level of enterprises operated well is in a leading position, and the credit risk can reduce. While the enterprises operated poorly may face the risk of losses or failures. This paper will discuss from two perspectives of the production level and operating conditions of large power customers.

2.3. Financial Credit Risk

The financial situation of large power customers reflects the enterprise management level, will also affect the future development and the competitiveness of the enterprise. The financial risk can be divided into solvency risk, profitability risk and development capacity risk.

2.4. Transaction Credit Risk

The transaction credit risk can be divided into direct trade credit risk and demand side response to the credit risk according to the type of market transactions that large power customers participate in.

3. Credit Risk Evaluation Method for Large Power Customers

3.1. Determine Index Score by Interval Number Method

3.1.1. Dimensionless Treatment for Interval Numbers. Dimensionless treatment is to eliminate the influence of dimensional difference among different indicators on evaluation.

$$Y_{Max,ij}^{P} = X_{ij}^{P} / \sum_{i=1}^{m} X_{ij}^{P}, Y_{Max,ij}^{Q} = X_{ij}^{Q} / \sum_{i=1}^{m} X_{ij}^{Q}, Y_{Min,ij}^{P} = \frac{1/X_{ij}^{P}}{\sum_{i=1}^{m} 1/X_{ij}^{P}}, Y_{Min,ij}^{Q} = \frac{1/X_{ij}^{Q}}{\sum_{i=1}^{m} 1/X_{ij}^{Q}} \quad (1)$$

3.1.2. Value Transformation for Interval Numbers. Supposing that interval numbers are subordinated to normal distribution and exchanged into evaluation value by standardization.

$$u_{ij} = \frac{X_{ij}^{P} + X_{ij}^{Q}}{2}, \sigma_{ij} = \frac{X_{ij}^{Q} - X_{ij}^{P}}{6}, \sigma_{ij} = u_{ij} + \lambda \times \left(1 - \sigma_{ij}\right) \quad (2)$$

3.2. Credit Evaluation Process Based on AHP and Entropy Weight Method (EWM)

3.2.1. Determine Objective Weight by Entropy Weight Method (EWM).

(1) Calculate the proximity between each large power customer index and the optimal value.

$$x_{ij} = x_{ij} / x_{j}^{*}, x_{j}^{*} = Max\left(x_{ij}\right), x_{ij} = x_{j}^{*} / x_{ij}, x_{j}^{*} = Max\left(x_{ij}\right), d_{ij} = \frac{x_{ij}}{\sum_{j=1}^{m} \sum_{i=1}^{n} x_{ij}} \quad (3)$$

(2) Calculate importance entropy of index.

$$E_{ij} = -d_{ij} \times \ln\left(d_{ij}\right), e_{ij} = -\frac{d_{ij}}{\sum_{j=1}^{m} d_{ij}} \times \ln\left(d_{ij} / \sum_{j=1}^{m} d_{ij}\right), e_{j} = \frac{1}{\ln(m)} \times \sum_{j=1}^{m} e_{ij} \quad (4)$$

(3) Calculate weight of index.

$$u_{j} = \left(1 - e_{j}\right) / \left(n - \sum_{j=1}^{n} e_{j}\right) \quad (5)$$

3.2.2. Determine Objective Weight by AHP.

(1) According to the experience of experts and credit rating workers' opinion, criterion indexes are ranked by importance in pairs, which form a judgment matrix.

(2) Calculate weight feature vector.

$$M_{i} = \prod_{j=1}^{n} x_{ij}, \overline{W_{i}} = \sqrt[n]{m_{i}}, W_{i} = \overline{W_{i}} / \sum_{j=1}^{n} \overline{W_{i}} \quad (6)$$

3.2.3. Calculate Comprehensive Evaluation Value.

Calculate maximum eigenvalue of matrix and get comprehensive evaluation value.

$$\varphi_{Int} = 0.5\times\varpi_{AHP} + 0.5\times\varpi_{EWM}, \chi = D\times\varphi_{Int} \tag{7}$$

4. Credit Rating of Large Power Customers

According to the evaluation process above, subjective and objective weight are calculated to get integrated weight of large power customers, as shown in **Table 2**.

Table 2. Interval Number of Large Power Customer Evaluation Indexes.

Target Layer	Weight	Primary Factor Layer	Weight	AHP	EWM	Integrated Weight
Credit Risk of External Environment	0.0776	Government Support	0.2275	0.0043	0.0000	0.0021
		Industrial Macro-Control Policies	0.4834	0.0091	0.0000	0.0046
		Fluctuation of Supply and Demand	1.0798	0.0204	0.0425	0.0314
		Consolidated Price Increase	2.3280	0.0437	0.0425	0.0431
Credit Risk of Production and Operation	0.2010	Market Share	0.4396	0.0211	0.0450	0.0331
		Corporate Social Impact	0.9356	0.0435	0.0451	0.0443
		Product Qualification Rate	1.4270	0.0681	0.0422	0.0552
		Safety Accident Rate	1.4270	0.0681	0.0425	0.0553
Credit Risk of Financial	0.2010	Return on Assets	0.4956	0.0142	0.0487	0.0315
		Net Profit Margin on Sales	0.8806	0.0251	0.0379	0.0315
		Cost Profit Margin	1.5491	0.0444	0.0568	0.0506
		Asset Liability Ratio	0.4956	0.0142	0.0476	0.0309
		Current Ratio	0.4596	0.0142	0.0386	0.264
		Main Business Revenue Growth Rate	1.5491	0.0444	0.0841	0.0643
		Net Profit Growth Rate	1.5491	0.0444	0.0841	0.0643
Credit Risk of Transaction	0.5205	Full Payment Rate	0.9776	0.0635	0.0425	0.0530
		Time Contribution Ratio	0.5253	0.0341	0.0446	0.0394
		Contracting Rate of Electricity Sales Contract	0.9776	0.0635	0.0429	0.0532
		Actual Power Completion Rate	1.7587	0.1139	0.0425	0.0782
		Contract Performance Rate	1.7587	0.1139	0.0425	0.0782
		Interruptible Load Contract Performance Rate	0.9776	0.0635	0.0417	0.0526
		Load Control Coordination	0.5253	0.0341	0.0448	0.0394
		Electricity Check Fit	0.5253	0.0341	0.0412	0.0376

As **Figure 1** shown, comprehensive evaluation value of large power customer 3 is the highest, followed by large power customer 2. Credit rating of these two corporates is at a high level, which indicates that they have a certain reputation in the industry with good business development, high debt credit, strong resistance to external environmental impact and low default risk.

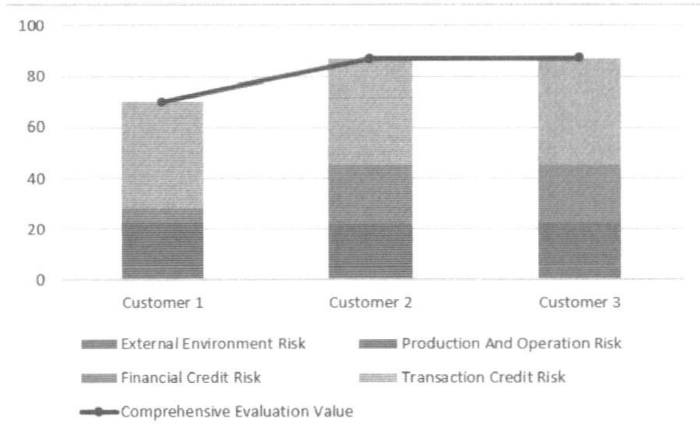

Figure 1. Evaluation Value of Large Power Customer Credit.

5. Conclusion

With the development of power system reform, credit problem in power industry will be more complicated. In this paper, a set of credit risk evaluation index and evaluation model based on comprehensive evaluation are constructed. As the results of the numerical examples shown, credit risk of large power customers is closely related to external environment risk, production and operation risk, financial credit risk and transaction credit risk. Also, we can compare and analyze the credit rating of different enterprises and the indicators to be improved. The model established in this paper provides a quantitative reference for the credit prevention of power market transactions.

References

[1] Hao Yang. Research on Risk Management of Electricity Power Supply Enterprises in China. Beijing, North China Electric Power University, 2014.
[2] Chen Yanchao. Study of the Credit Evaluation of Key Customer and the Risk Management of the Power Company. Beijing, North China Electric Power University, 2015.
[3] Luo Yifan. Research on Electricity Customer's Credit Rating and Anti-stealing Power Considering Distributed Generation. Shanghai, Shanghai Jiao Tong University, 2015.
[4] Streimikiene D, Ciegis R, and Grundey D. Energy Indicators for Sustainable Development in Baltic States. Renewable & Sustainable Energy Reviews, vol. 5, Dec 2007, pp. 877-893.
[5] Niu Xiaomei, and Zhang Yinling. Assessment of Power Customer Credit Risk Based on Analytic Hierarchy Process. Computer Simulation, vol. 5, 2011, pp. 333-336.
[6] Huang Yuansheng, and Shi Xiufen. Evaluation of Electricity Customers' Credit Risk based on Entropy Method and MADM. Technoeconomics & Management Research, vol. 2, 2010, pp. 24-27.
[7] Lu Jianchang, Jiang Wei, and Liu Na. Study on Credit Evaluation of Power Consumers Based on Improved Entropy. Journal of Electric Power, vol. 5, 2008, pp. 368-370.

Acknowledgment

This work is sponsored and supported by National Power Grid Corp science and technology project *Research Application of Key Technologies for Power Grid Planning, Investment and Purchasing Power Optimization* (Contract No. SGZJJY00SJJS1600031).

Research on the operation control strategy of the cooling ceiling combined with fresh air system

Tao Huang[1] and Hao Li [2]

[1]Institute of Building Environment and Energy, China Academy of Building Research, Beijing, 100018, China
[2]School of Mechanical Engineering, Tianjin University, Tianjin, China

E-mail:huangtao_ht@126.com

Abstract. The cooling ceiling combined with independent fresh air system was built by TRNSYS. And the cooling effects of the air conditioning system of an office in Beijing in a summer typical day were simulated. Based on the "variable temperature" control strategy, the operation strategy of "variable air volume auxiliary adjustment" was put forward. The variation of the indoor temperature, the indoor humidity, the temperature of supplying water and the temperature of returning water were simulated under the two control strategies. The energy consumption of system during the whole summer was compared by utilizing the two control strategies, and the indoor thermal comfort was analyzed. The optimal control strategy was proposed under the condition that the condensation on the surface of the cooling ceiling is not occurred and the indoor thermal comfort is satisfied.

1. Introduction

The advantage of cooling ceiling system is different from traditional air-conditioning system for it has more energy saving potential, and can make the indoor environment more comfortable and healthy. Therefore, it attracts extensive researches and attentions in recent years [1]. But in practical engineering applications, especially in hot and humid areas in summer, the phenomenon on the condensation of radiant roof often occurs because of the improper control of the system. In addition, due to the cooling capacity is small and the room temperature is difficult to control, the popularization and application of cooling ceiling system are hindered seriously[2]. In the cooling ceiling combined with independent fresh air system, the fresh air is sent into the room after the process of dehumidification and cooing, the fresh air not only can undertake indoor wet load and a certain extent indoor cooling load, but also meet the needs of the indoor fresh air. The form of energy exchange between the radiation surface and the indoor personnel is radiation, which greatly improves the indoor comfort, and can create an indoor environment with low wind speed and a uniform temperature field.
The air volume of the cooing ceiling system is much lower than that of the conventional air conditioning system, and the energy consumption of the transportation system can be greatly reduced [3].

In order to give full play to the advantages of the cooling ceiling air-conditioning system, and provide a valuable reference for its popularization, the cooling ceiling combined with independent fresh air system was built by TRNSYS, in where the energy matching characteristics between the key components in the cooling system and cooling load demand are considered.

2. Building overview

The research object is an office building facing southwest: the floor area is 185m^2, concrete wall thickness is 200mm, and the external insulation thickness is 55mm. The building model is shown in figure 1. The outside windows are double-layer hollow Aluminum Alloy broken bridge windows and the window thermal conductivity is 2.2W/ (m^2·K), the building height is 3.6meters, and the actual ceiling height is 2.8meters. The power density of indoor lighting is 10W/m^2, and the computer equipment power is 18W/m^2. The fresh air volume is 1000m^3/h, and the office containing 22 people.

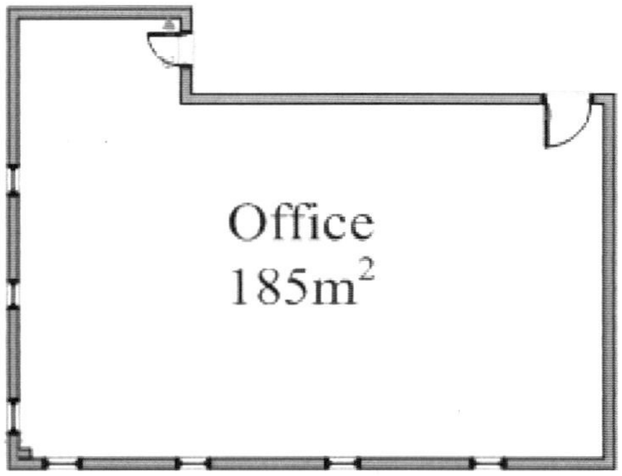

Figure 1. Building model

The ceiling of the room is 20 cm below the floor slab, and the ceiling is made of metal plate. This kind of metal plate ceiling is very suitable for office buildings because of its small thermal resistance, quick response and the integration of cooling and decoration.

The building model is built by TRNSYS, and the cold load of the room is simulated by using the typical year weather data. The result is shown in Figure 2. The maximum cold load of the room is 13.35 kW, which is the basis of system design and equipment selection.

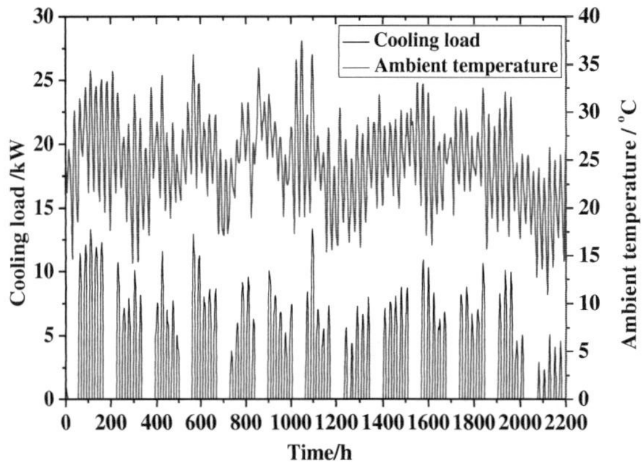

Figure 2. Building cooling load

3. System design and control strategy

3.1. System constitution

The cooling ceiling combined with independent fresh air system is a kind of temperature and humidity independent control air conditioning system. The schematic diagram is shown in Figure3. The chiller is an air source heat pump with the cooling capacity 14kW. The supply and return water temperature are 7℃and 12℃respectively and the running time of system is 07:00~20:00. The main water of the chiller is carried to an independent fresh air unit, in which outdoor fresh air is cooled and dehumidified. The water of the independent fresh air unit is supplied into the radiation coils after the flow is distributed through the distributary three-way valve and mixing valve.

IOP Conf. Series: Earth and Environmental Science **127** (2018) 012018 doi:10.1088/1755-1315/127/1/012018

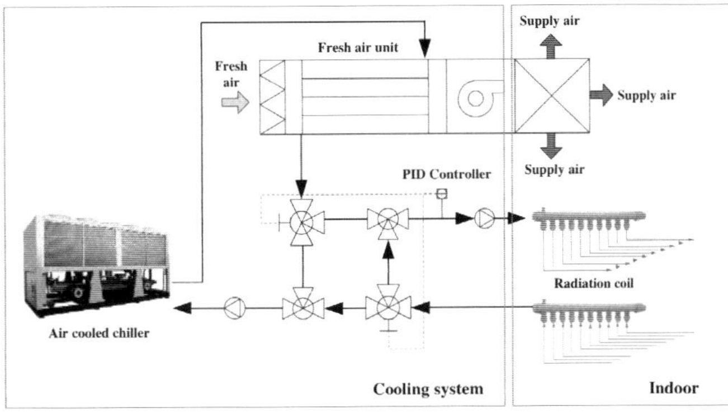

Figure 3. The schematic diagram

3.2. Control strategy
The operating mode of the system is related to the energy consumption of the system, and also directly affect the indoor thermal comfort. At the same time, in order to prevent the phenomenon of condensation, the water temperature is must higher than the indoor dew point temperature. Generally, cooling ceiling system is not suitable for flow regulation because of small temperature difference and the operation of high flow. The relevant research results show that the influence of regulating flow rate on cooling capacity is little. At the same time the change of flow rate will lead to hydraulic imbalance and exhaust difficulties[4]. Therefore, the flow control is not involved in this study.

In this study, the PID controller is used to monitor the supply water temperature. Under the condition that the supply water flow is constant, the proportion of the flow distribution is adjusted, so that the supply water temperature is higher than the indoor dew point temperature 1℃, so as to avoid the condensation.

The starting and stopping of radiation water pump is according to the change of indoor temperature. The operation mode is: when the indoor temperature is higher than 25℃, the radiation water pump is opened, and the radiation water pump is closed when the temperature is lower than 23℃.

4. Model constructing
TRNSYS software is used to simulate the cooling ceiling combined with independent fresh air system, and the operation and control process is shown in Figure4.

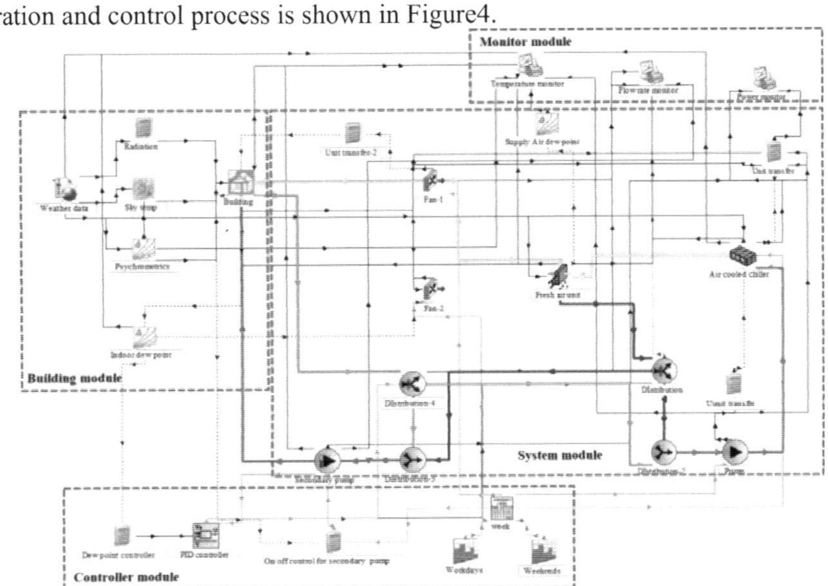

Figure 4. Model composition for TRNSYS

The experimental system built on the TRNSYS platform include the building module, refrigeration system module, control module and monitoring module.

The simulation room is established with the Type65b module in TRNSYS[5]. The indoor initial setting point parameter is 30℃, and the relative humidity is 50%. The structural parameters of the radiant ceiling plate are realized by setting "chilled ceiling" in the "wall type manager" option. The independent fresh air unit is set through Type508h module. In addition, the Type665 module which is air cooled chiller is used in the refrigeration unit, the Type33e module is used to calculate indoor dew point temperature and radiation pump is established with multi-stage transmission pump Type110 module[6].

5. Analysis of simulated result

5.1. "Variable temperature" control strategy

The change curve of indoor temperature and other related temperature parameter were shown in Figure 5, which the "variable temperature" control strategy was employed. In order to ensure that the indoor environment can meet the requirements of work, so the system needs to be opened one hour in advance.

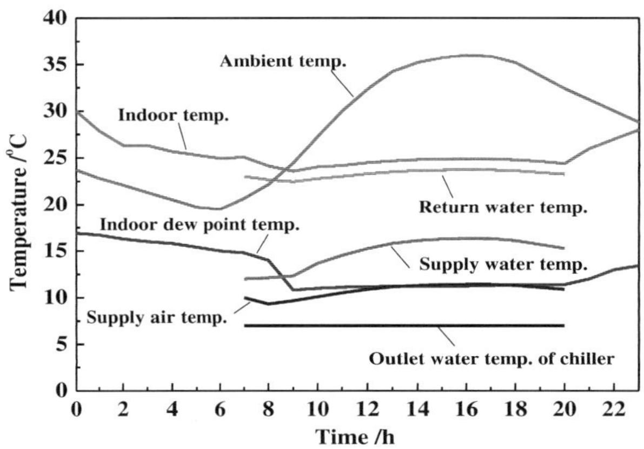

Figure 5. Change of relative parameters of "variable temperature" control strategy

From the simulation results, we can see that during the first one hour which the radiation pump is opened, the supply water temperature is lower than the dew point temperature, which will cause the radiation ceiling to dew. The indoor temperature can be stable at 24℃ from 8:00 to 20:00. In the initial stage of the operating system, due to the hysteresis of PID control, the system runs unstably and the supply water temperature of the radiation water pump is lower than 12℃. In the latter stage of the operating system, the temperature of the supply water is 15~16℃. The temperature of the backwater is stabled at 22~23 ℃.

5.2. Variable air volume auxiliary adjustment

In order to solve the radiant ceilings dewing phenomenon in the early stage of running system, the operation strategy of "variable air volume auxiliary adjustment" was put forward. That is, in the initial of the system operation only the fresh air system is opened for a period of time, and the circulating air volume is increased, so that the indoor temperature and humidity drops rapidly. AT the working time, the fresh air system is returned to the normal state, and the radiation pump is opened.

The change curve of indoor temperature and other related temperature parameter were shown in Figure 6, which the "variable air volume auxiliary adjustment" control strategy was employed. From 7:00 to 8:00 the supply water temperature is always higher than the dew point temperature, which will not cause the radiation ceiling to dew.

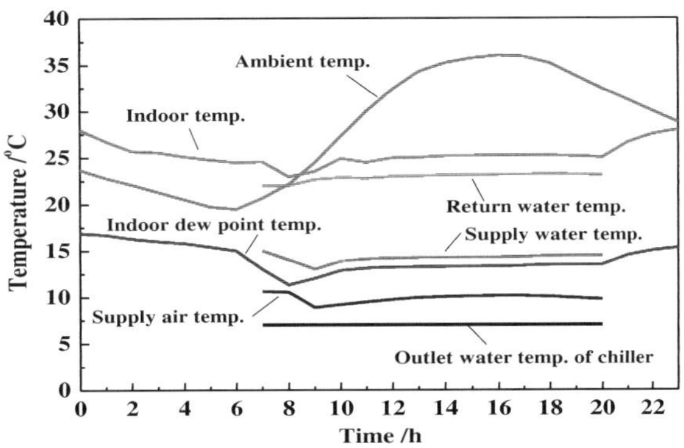

Figure 6. Change of relative parameters of "variable air volume auxiliary adjustment"

The indoor thermal comfort was simulated and the result is shown in Figure 7. During the office work time, 08:00~20:00, the indoor thermal comfort fully meet the ISO7730 recommended value of PMV-PPD indicators, that the PMV value is about 0.1 and PPD is about 5%.

5.3. Energy consumption of system

The total time of the simulation for the cooling season is 2208h and daily operation time is 07:00~20:00. The energy consumption statistics of chiller, water pump and fans are shown in table1 by using the two control strategies.

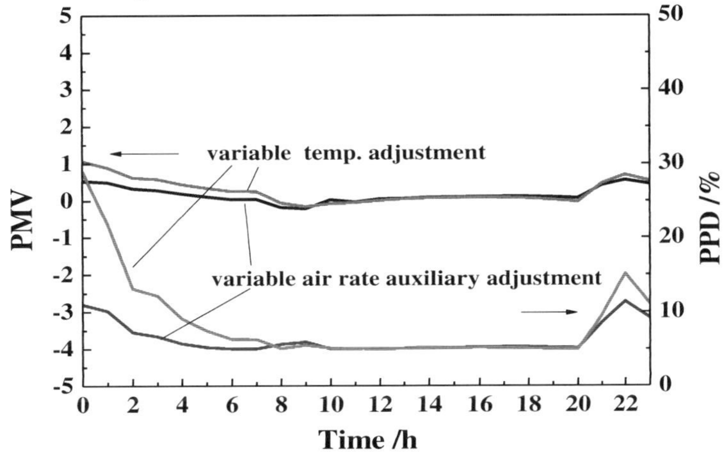

Figure 7. Indoor thermal comfort

From the analysis of energy consumption, we can see that the energy consumption level of the two control methods is quite equal. The energy consumption of "variable air volume auxiliary adjustment" control mode is slightly higher, mainly due to the indoor load is served by the fresh air unit in the initial stage of the system operation every day. At the same time, the backwater temperature of chiller is relatively low, which results in the decrease of the COP and the increase of the energy consumption of the chiller.

Table 1. Statistical results of energy consumption

Energy consumption kW·h	Variable temperature control	Variable air volume auxiliary adjustment
chiller	1297.98	1369.54
Primary circulation pump	66.50	66.54
fan	27.23	35.46.
Radiation water pump	42.72	37.23
Total	**1434.43**	**1473.31**

6. Conclusion

From the point of system control strategy, the cooling ceiling combined with independent fresh air system was built by TRNSYS. Based on the strategy of variable water temperature control, the operation strategy of variable air volume auxiliary adjustment was put forwarded in this paper.

(1) Through comparing the two regulation methods, it was found that the indoor thermal comfort with PMV-PPD meets the requirements during the period of 08:00~20:00. The results by using the two regulation methods are close to each other, which values of PMV are about 0.1 and values of PPD are about 5%.

(2) The "variable air volume auxiliary adjustment" eliminates the possibility of condensation on the surface of radiant ceiling at the initial stage of system operation.

(3) The energy consumption of "variable air volume auxiliary adjustment" control mode is slightly higher in the whole cooling season. This is because the returning water temperature of chiller is relatively low, which results in the decrease of COP of the chiller in the early stage of system operation every day.

7. References

[1] Wang Zijie, "Low temperature radiant heating and radiant cooling", beijing, China Machine Press, 2004.
[2] Yang Hailong and Wang Houhua, "Inquiry on control system of radiant suspended ceiling for protecting condensation" Shanxi Architecture, 2011, 37(33), pp.124-125.
[3] Bao Feng, "Application of radiation cooling ceiling combined with mechanical ventilation", Huazhong University of Science and Technology, 2009.
[4] Ding Yan, "Research on air supply type and control strategy of radiant cooling system with independent fresh air", Tianjin University, 2012.
[5] LU Shikui, LV Yan, ZHANG Xiao-song and CHEN Jie,Thermal Comfort Assessmentand Energy Consumption Analysis about the System of Mechanical Ventilation Combined with Radiant Cooling", *Building Energy&Environment*, 2010, 29(3),PP:60-62.
[6] Zhao Yu, Yuan Dongli and XieFei, "Regulation performance of cooling ceiling air conditioning system", HV&AC, 2014, 44(2), pp:65-68.

Acknowledgment

This work was financially supported by National Natural Science Foundation (No.51408576).

IOP Conf. Series: Earth and Environmental Science **127** (2018) 012019 doi:10.1088/1755-1315/127/1/012019

Effect of Policy Analysis on Indonesia's Maritime Cluster Development Using System Dynamics Modeling

A Nursyamsi, A O Moeis and Komarudin

Systems Engineering, Modeling, and Simulatioin Laboratory, Industrial Engineering Department, Faculty of Engineering, Universitas Indonesia, Depok, Indonesia

aditya.nursyamsi_ti2013@yahoo.com

Abstract. As an archipelago with two third of its territory consist of water, Indonesia should address more attention to its maritime industry development. One of the catalyst to fasten the maritime industry growth is by developing a maritime cluster. The purpose of this research is to gain understanding of the effect if Indonesia implement maritime cluster policy to the growth of maritime economic and its role to enhance the maritime cluster performance, hence enhancing Indonesia's maritime industry as well. The result of the constructed system dynamic model simulation shows that with the effect of maritime cluster, the growth of employment rate and maritime economic is much bigger that the business as usual case exponentially. The result implies that the government should act fast to form a legitimate cluster maritime organizer institution so that there will be a synergize, sustainable, and positive maritime cluster environment that will benefit the performance of Indonesia's maritime industry.

1. Introduction

As a large archipelago country with two thirds of its territory consist of water, Indonesia is known as maritime country. Therefore, it is not surprising that Indonesia should pay more attention to the maritime industry to support Indonesia's economic growth. Indonesia has various maritime industry sectors, ranging from fishery industry, shipyard, to logistics transportation.

Some maritime industries in Indonesia show good performance. One of these is the fishing industry, which ranks 4th in the world [1]. The result is accompanied by the development of catching fish in Indonesia which has the tendency to increase from year to year [2]. The logistics transportation industry in Indonesia is also good when viewed from the volume of loading and unloading of goods in the port that has a tendency to rise from year to year [3].

However, Indonesia's Logistics Performance Index (LPI) remains at number 63 [3], far below other countries, including neighbouring countries such as Singapore and Malaysia. This shows that Indonesia still insufficient in exploiting its maritime potential. In fact, according to Deputy Minister of Industry, Mr. Alex S.W Retraubun, "The maritime issues of the world exist in Indonesia. However, Indonesia's maritime industry has not been very good because there is no government support in terms of fiscal policy towards the development of technology, components, and Human Resources (HR)".

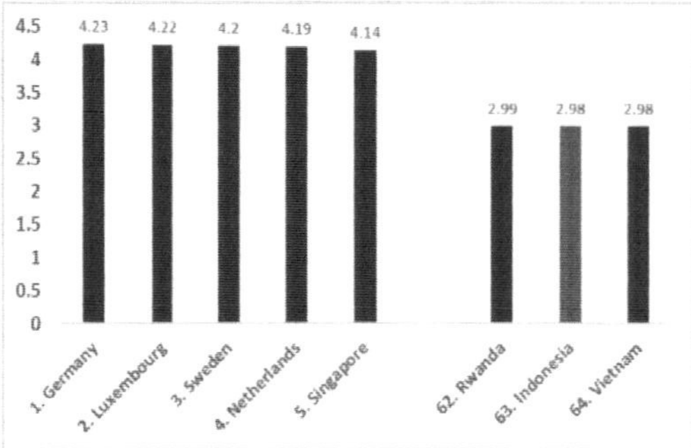

Figure 1. Logistic Performance Index

As a large maritime country, Indonesia has not been able to maximize the potential of its maritime industry. This can be seen from the contribution of the maritime sector in Indonesia that does not reach 20% of the total value of Gross Domestic Product (GDP) [2]. Whereas in other fellow maritime countries, their maritime industry can contribute up to 50% of its total GDP [3]. Without research related to maritime clusters, Indonesia will lose momentum to accelerate the of its maritime industry. Therefore, a maritime cluster research that can integrate the role of government and actors in the maritime industry to adopt a strategic policy to ensure the progress of Indonesia's maritime industry in the future.

The purpose of this study is to gain an understanding of the effects that can occur if Indonesia implements a maritime cluster policy on maritime economic growth and its role to maximize Indonesia's maritime industry potential, and, in the long run, can also advance Indonesia's economy. He output resulting from this research is a system dynamics simulation model, the comparison of simulation results when Indonesia implemented maritime cluster policy and not, and the result of policy analysis which can be the main recommendation to be taken or implemented by the government.

2. Theoritical Background

2.1. Indonesia's Profile

Indonesia is undoubtedly one of the world's largest countries. Based on the total area (land and sea), Indonesia is ranked 7th of the world with total area of 5,193,250 km2, where total land of Indonesia reaches 1,811,569 km2 and total of Indonesia ocean reach 3,288,681 km2. Geographically, Indonesia lies between two continents, namely Asia and Australia, and between the two oceans, the Pacific Ocean and the Indian Ocean. Because of its strategic location, Indonesia is famous in the eyes of the world as an International trade route.

As a developing country, Indonesia has a strong population and economic power. With a total population of over 250 million people, Indonesia ranks 4th in the world as the country with the largest population and with a fairly stable population growth rate of 1.3% per year [2]. Indonesia has a labour force participation rate of 65.8% of the total population or about 168 million people. However, the unemployment rate in Indonesia is still relatively high in the range of 6.2% or about 15 million people. This indicates that Indonesia still lacks industry or employment to absorb unemployment.

Indonesia has a pretty good and stable economic profile. One of the economic parameters that can be taken is total Gross Domestic Product (GDP). Indonesia's GDP in 2015 reached more than 11 trillion rupiahs, with economic growth rate reaching an average of 4.8% per year [2]. However, despite having so much GDP, the value of GDP per capita in Indonesia is relatively small compared to other ASEAN countries at 45.2 million rupiahs, while neighbouring countries such as Singapore can reach 52 thousand USD or around 670 million rupiahs [3].

As a country with an area of the ocean more than two-thirds of the entire region, it is not surprising that Indonesia should depend on its maritime power. As per the direction of the President, it is time for Indonesia to return to the sea, again making the sea as the power of the nation, so that Indonesia can become the World Maritime Fulcrum [4] – [7]. The vision of Indonesia to be a World Maritime

Fulcrum based on historical reviews, geo-strategic location, and the desire for the existence of the nation should be increasingly taken into account at regional and global level. This vision is a long-term step that requires consistency of policy, cooperation, and sincerity to achieve it. To achieve the vision to be World Maritime Fulcrum, several important aspects of maritime possessed by the nation are clearly mapped out their strengths and weaknesses, as well as known challenges. Furthermore, the President conveyed 5 (five) pillars of Maritime Fulcrum Development [7] in his speech at East Asian Summit 2014 covering:

a) Develop maritime culture
b) Maintain and manage marine resources
c) Development of maritime infrastructure and connectivity
d) Strengthening maritime diplomacy
e) Build a maritime defense force

2.2. Industry Cluster

Cluster concepts have long been a hot topic of research discussed in major countries in terms of economic analysis. Research related to the analysis of economic relations to clusters began to grow rapidly since Michael E. Porter conducted related research in his writings "The Competitive Advantage of Nations" [8]. Porter developed the theoretical basis for cluster concepts as a competitive advantage for a country. According to Porter, industrial clusters are a cooperative group with an adjacent geographic scope that includes suppliers, users, supporting industries, government and support institutions such as universities. In a later study, Porter reiterated his understanding of industry clusters, a group of interconnected firms together with related institutions in a specific field, adjacent geographically, linked by similarity and complementarity [9]. Morosini then provides his understanding of the industrial cluster as a socio-economic entity categorized by the social community of persons and populations of economic agents residing within the same geographical environment [10]. Based on the description, can be drawn character from a cluster concept as follows [11]:

- Regional economic activity is located at various levels: community, geographic area, global
- Limited to specific industries
- Covers both vertical relationships such as supplier-factory-dealers-users as well as horizontal relationships lilke other sectors in the same industry
- The company has the same or inteconnected business areas
- Each company competes with each other but through their own specialization contributes to the development of clusters
- The company's closeness generates social relationships and a sense of trust
- Common infratructure used to innovate due to strong knowledge transfer and supported by universities and research centers

While there is no definite theory of clusters yet, the advantages of a cluster have been observed, as well as the main reasons that such research is often the focus of research [8], [10], [12]:

- High concentrations of firms generate an icreased market and provide opportunities to get more buyers
- Reduced transportation costs
- Easier access to resources
- Opportunities for new companies that see potential in a cluster environment
- Offers a high level of specialization for products and services within a cluster development
- A more competitive environment serves as a better motivation
- Stronger cooperation among cluster members; Intimacy between companies increases trust between companies and improves communication
- Concentration of companies engaged in the same field provides an expert field of labor in the field
- Better access to reliable employees

In the case of cluster development, the government should have an important role. Governments have an important role to play in facilitating the development and enhancement of a cluster quality. According to Norman and Venables, government policy also has an important role in cluster development. Without a government policy, the number of industrial clusters will be too much, and each cluster will become too narrow, which in turn will reduce the output that should be generated [13]. Governments can encourage cluster development in a variety of ways, one of which is to remove barriers and reduce constraints, including human resources, infrastructure and related regulations.

2.3. Maritime Cluster

Given the relationships created through a cluster, the maritime industry recognizes that there is a great advantage of agglomeration or adjacent locations within a region [14] – [15]. These advantages are in the form of knowledge dissemination, the availability of sufficient labor, a strong innovation culture, and a supportive market.

In developing a maritime cluster, the important points to note one of which is what industry or company deserves to be included. Each country can define what industry or company it wants to include in establishing a maritime cluster depending on the conditions of their country. Thomas Engelke defines sectors in the maritime sector to 9 sectors (Figure 2) [16]. The ninth should be strengthened by using an integrated approach to create a strong maritime sector within a country.

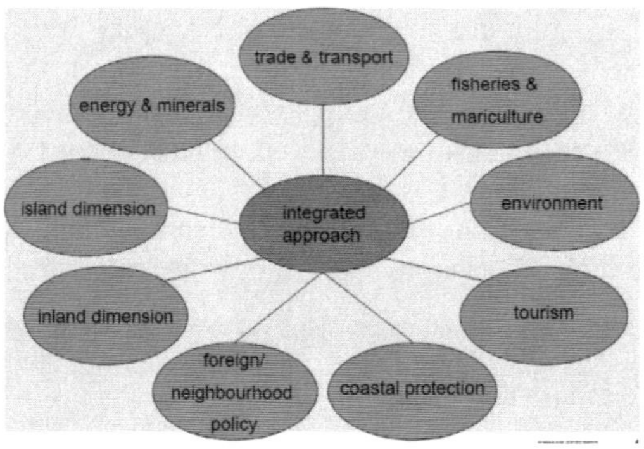

Figure 2. Maritime Cluster.

In the development of a maritime industry system, known as multi-level perspective. As can be seen in Figure 3, the perspective in the development of the maritime industry can be divided into 3 levels. First, the macro level in the form of maritime cluster research with coverage of Indonesia. Next there is a intermediate level that can address one node in the maritime cluster, for example a port cluster research. Then there is a micro level that can discuss about the company's performance analysis in the maritime industry.

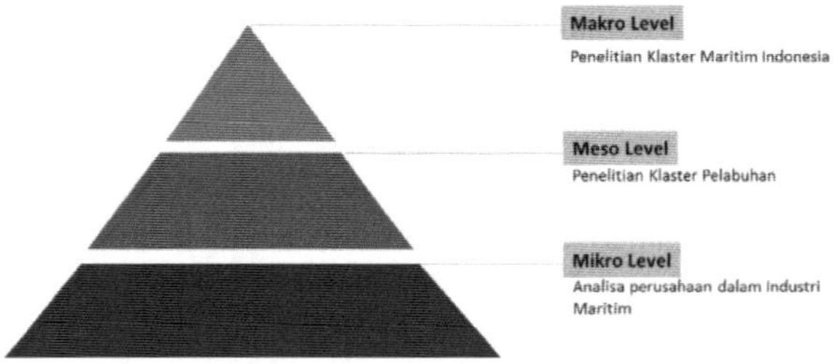

Figure 3. Multi-level perspective on Maritime Industry Research.

The concept of whether or not a cluster is strong is determined by the type and number of its maritime sector companies [15]. The more types of companies in a cluster, the greater the synergy and strength generated by the cluster. The relationship between cluster completeness in terms of type of company and cluster strength can be seen in Figure 4. It is seen that there is a non-linear relationship in the relationship.

Assessment indicators for a cluster can be measured qualitatively and quantitatively. Quantitative indicators such as economic growth, GDP, number of labor, number of companies in clusters, and so forth. As for the qualitative indicators for example are trust and coordination, the general impact on the economy, as well as the relationships created between the company with other companies, research institutions or education, as well as policy makers. In a cluster, apply the concept that the more players, the stronger the cluster. The concept applies to all indicators.

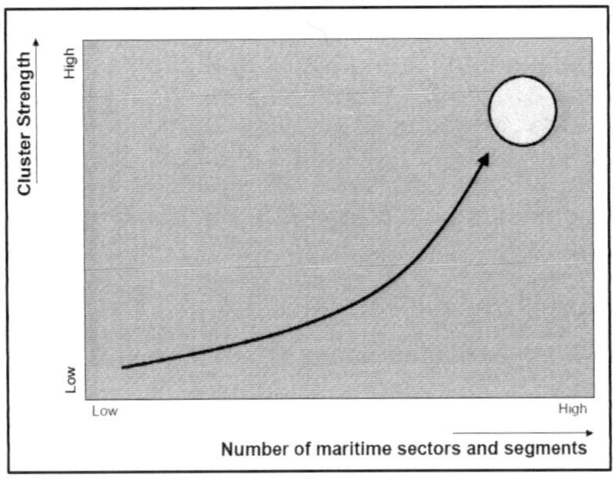

Figure 4. Reationship between number of sectors and Cluster Strength

In order to build a good maritime cluster, model makers need to understand what factors can strengthen the cluster. Wijnolst has defined 7 factors that can make a cluster successful. These are the following factors [14] – [15]:

- Define clusters clearly
- Create an industry policy to support the cluster
- Strengthen the demand-pull sector
- Overseeing the company's operations so that there is no element of fraud
- Promoting export and internationalization
- Strengthen innovation and R&D
- Strengthen the quality of education and manpower

Policy is one of the factors that can support the quality of maritime clusters. With the right government and policy support, a country has a great opportunity to strengthen its maritime cluster. Engelke provides several examples of policies that can be taken to support the development of a maritime cluster [16]. These policies are governance, education, public awareness, cluster, regional planning, maritime technology, and research. The policy should be made with an integrated approach so that it can create the correct policy package as needed.

2.4. System Dynamics

In 1960, Jay W. Forrester, a researcher from MIT published the results of his research he has done for about 4 years about dynamic system modelling. According to Richardson, the dynamic system approach itself is actually a combination of several approaches, computing technology, computer simulation, strategic decision making, and understanding of feedback in a complex system [17].

There are four basic concepts that are important to keep in mind when constructing a dynamic system model in order to clearly illustrate the structure and behaviour of a complex system, which are the nature of a closed system (endogenous), the effect of feedback, there are basic variables of levels and rates, and there are success parameters from model development.

The main purpose of developing a dynamic system model is to study, recognize, and understand the structure of a complex system so that it can then understand the role of policy and the delay of a decision in influencing the behaviour of the system. The main focus of the system dynamics methodology is to gain an understanding of a system, so that problem solving steps provide feedback on system understanding.

2.5. *Uncertainty in Model*

The concept of uncertainty can be interpreted in general as a limited knowledge of the future, past, or events in the present. Regarding policy-making issues, the concept of uncertainty is also influenced by subjectivity, as it relates to satisfaction of existing knowledge, which is influenced by the values and perspectives of policy makers and the various actors involved in the policy-making process, and Influenced also by the choice of decisions available to them. Uncertainty does not necessarily mean that lack of knowledge, Funtowicz and Rayetz describe uncertainty as a situation where there is insufficient information. Inadequate information is in the form of inappropriateness, unreliable sources, and bordering on ignorance. Walker provides four ways to address the issue of deep uncertainty in policy making [18]:

- *Resistance*: Plan for all possible worst case in the future
- *Resilience*: Whatever happens in the future, be sure to have policies that can neutralize the situation
- *Static robustness*: Implement policies that can work in different types of circumstances
- *Adaptive robustness*: Prepare to change policy, if conditions change

3. Methodology

The study was conducted using the principle of continuous iteration. The purpose of iteration process is that each step requires a feedback, which then the feedback will be implemented in the following process as well. Research is focused on developing a system dynamics model and observing at the effects of policies that can occur. Reference modes and system diagrams are formed as a basis for developing the dynamic systems of this research.

3.1. *Reference Mode*

In this study, the expected model behaviour is an assumption because the author will model a system for the future. The indicators of behaviour that are seen are the number of maritime workforce as well as the total maritime economic output driven by the industrial sector and the effects provided by the maritime cluster. According to Wijnolst, the relationship that will be created from the cluster against its effects on the economy and workforce is an exponentially growing relationship [14] – [15].

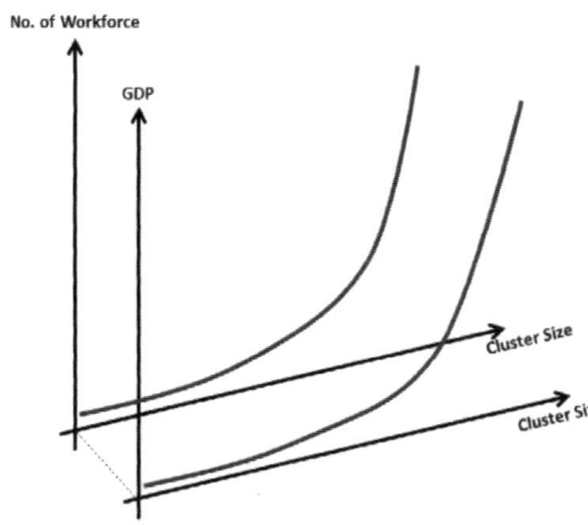

Figure 5. Reference Mode on Maritime Cluster

3.2. *System Diagram*

The system diagram is used to see the overall system of interrelationships between variables and the feedback that occurs on the context of the problems. A system diagram can show related actors, reciprocal relationships in the problem, and most importantly the output indicators of the model being created. Figure 6 illustrates the system diagram of this study.

IOP Conf. Series: Earth and Environmental Science **127** (2018) 012019 doi:10.1088/1755-1315/127/1/012019

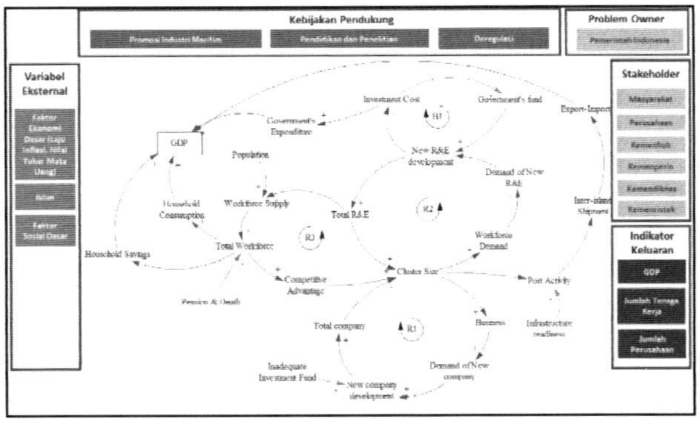

Figure 6. System Diagram

4. Result and Discussion

4.1. Model's Validation and Verification

Model verification is done regarding whether there are no more signs of error on each variable. The verification process is done directly within the software used in the model creation of Powersim Studio, where there are no more undefined variables and relationships.

Model validation is important to determine whether the model is appropriate or similar in representing the real system. However, to note is the statement that "all models are wrong". Validation test is done using 4 methods:

4.1.1. Error Integration Test. This test is performed by running simulations using 3 different timesteps; 1 year, 6 months, and 90 days. Based on the results obtained, there is no significant difference between the three timesteps as can be seen in Figure 7. This shows that model is valid based on the integration error test.

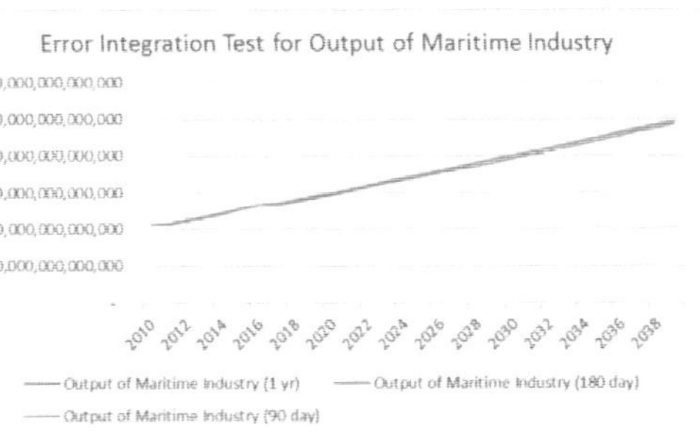

Figure 7. Validation Test - Error Integration

4.1.2. Extreme Condition Test. This test is done by changing the input value on the initial amount of maritime industry workforce. When the author changed the number to very extreme; from 6 million to 1 billion, the model shows that there will be no labor requirement so that there will be no increase in labor, only reduction. As can be seen in Figure 8, the trend of the labor force that occurs only decreases. Because the model still delivers sensible results when given extreme conditions, it can be said that this model is validated under extreme conditions tests.

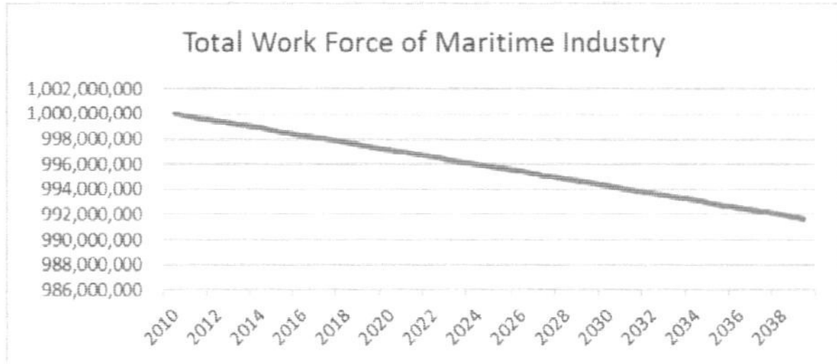

Figure 8. Extreme Condition Test on Workforce

4.1.3. Behaviour Reproduction Test. This test is done by comparing the behaviour between the value of the Maritime Industry output generated by the model with reality. It can be seen in Figure 9 that both of them have the same upward trend in both the simulation model and the reality with an error value of less than 5%.

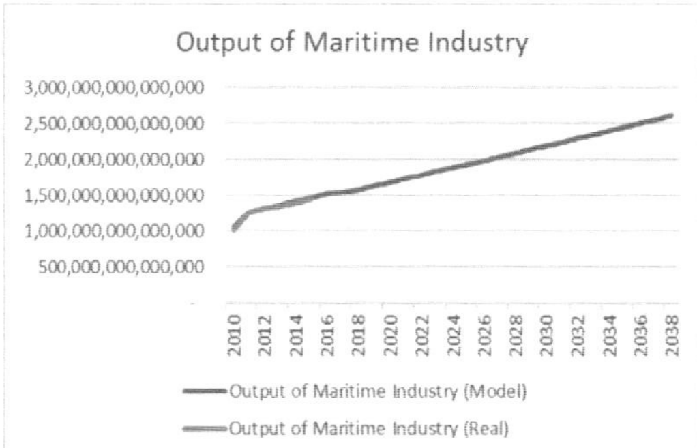

Figure 9. Behaviour Reproduction Test on Output of Maritime Industry

4.1.4. Structure Assessment Test. The created model already has a structure that is relevant to the system and concept of the existing problem. This can be seen from the suitability between the simulation model created on the causal loop diagram with the System Diagram as the framework.

4.2. Result Comparison Analysis of the Model

4.2.1. Growth Analysis of Small and Medium Enterprises following the Policy of the Maritime Cluster. The total number of small and medium enterprises between Business as Usual (BaU) and Maritime Cluster (MC) simulation models has a significant difference. The final total number of small and medium enterprises for BaU simulation model is 805 thousand business while for MC simulation model as much as 1.2 million business.

MC simulation model has a more positive growth rate compared to BaU simulation model. Both models also have different CAGR values, ~4% for BaU simulation model and ~5% for MC simulation model. This indicates that the simulation result using the MC simulation model is better than the BaU simulation model.

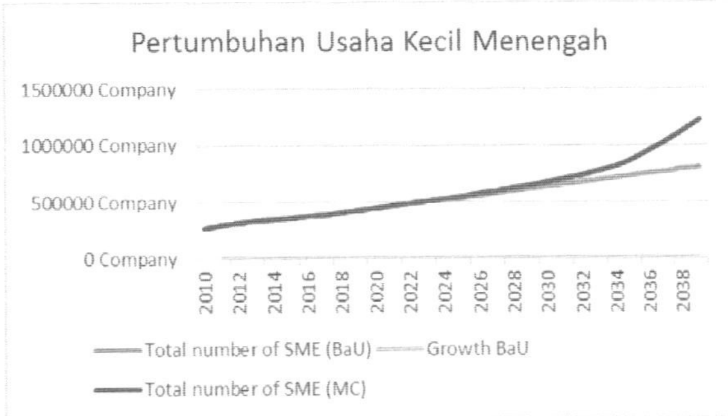

Figure 10 Growth Comparison on Small Medium Enterprises

4.2.2. Growth Analysis of Industry following the Policy of the Maritime Cluster. The total number of large Industries between BaU and MC simulation models has not significantly different, about 1000 companies. The final total number of large industry for BaU simulation model is 3451 companies while for MC simulation model as many as 4556 companies. However, because the contribution provided by large industries is greater than that of small and medium-sized businesses, even though the difference in quantities is only 1000 but the effect it provides on other indicators is considerable.

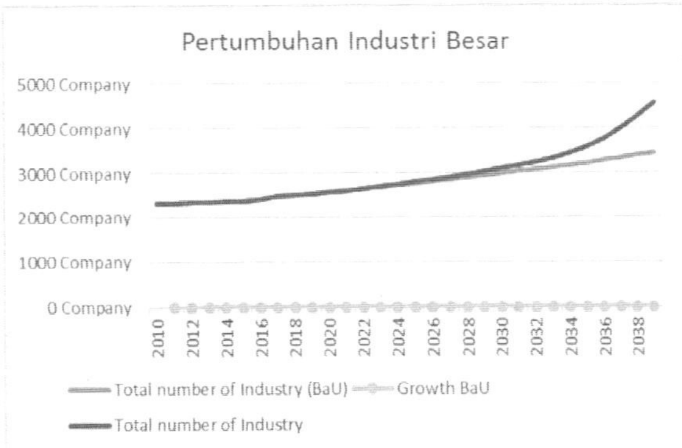

Figure 11 Growth Comparison on Industry

MC simulation model has a more positive growth rate compared to BaU simulation model. Both models also have different CAGR values, which is ~1% for BaU model and ~2% for MC simulation model. This indicates that the simulation result using the MC simulation model is better than the BaU model.

4.2.3. Growth Analysis of Maritime Industry Workforce following the Policy of the Maritime Cluster. MC model has a more positive growth rate compared to BaU model. Both models also have different CAGR values, ~4% for BaU simulation model and ~7% for MC simulation model. This indicates that the simulation result using the MC simulation model is better than the BaU simulation model.

Figure 12 Growth Comparison on Workforce

4.2.4. Growth Analysis of Maritime Industry Output following the Policy of the Maritime Cluster. MC model has a more positive growth rate compared to BaU model. Both models also have different CAGR values, ie ~3% for BaU simulation model and ~6% for MC simulation model. This indicates that the simulation result using the MC simulation model is better than the BaU simulation model.

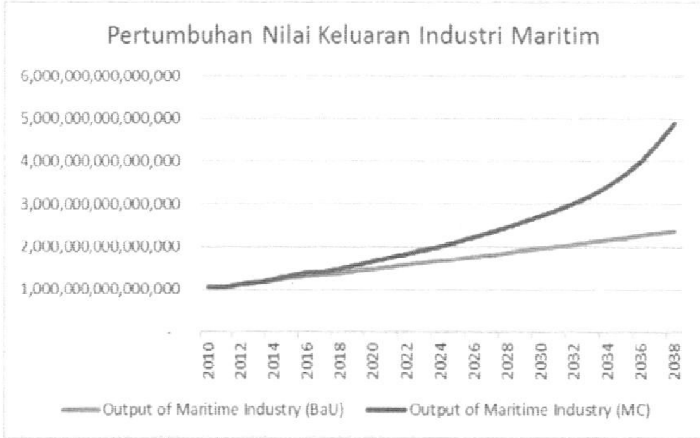

Figure 13. Growth Comparison on Output of Maritime Industry

5. Conclusion

5.1. Conclusion

Based on the analysis of the development process and the results of the model made, it can be concluded the things as follows:

- The model was developed to see the impact that could have occurred if Indonesia had a maritime cluster policy as a means of supporting the performance of the Maritime Industry.
- The model was developed based on the concept of maritime clusters that had already been developed in developed countries such as the European Union.
- Based on the results of the simulation, Indonesia will have higher economic and social value if it has implemented sustainable maritime cluster policy.
- Based on the workforce module simulation results, Indonesia will have a total workforce for maritime field with a CAGR of~7% when applying a maritime cluster policy compared to normal circumstances which only produce CAGR of ~4%
- Based on simulation result of economic module, Indonesia will have total value of Maritime Industry output with CAGR ~6% when applying maritime cluster policy compared with normal condition which only produce CAGR ~3%.
- Indonesia needs to establish a maritime cluster management institution in order to create a seamless, sustainable, and positive Maritime Cluster environment for the performance of the Maritime Industry in Indonesia.

5.2. Recommendation

Based on the conclusions of the research conducted, suggestions that can be given by author are:

- For further research can be explored further about the role of technology, culture of innovation, and environment towards the development of maritime clusters in Indonesia.
- Based on the results of the study, which still has limited data and information, it is expected that the Indonesian government may consider having a data center or conducting a special survey related to the statistics of the Maritime Industry in Indonesia in order to make it easier for researchers to conduct research related to maritime field in Indonesia.

6. References

[1] FAO Fisheries and Aquaculture Department. (2012). *The State of World Fisheries and Aquaculture 2012.* Rome: Food and Agriculture Organization of The United Nations

[2] Central Bureau of Statistics. (2016). *Indonesia Statistic 2016.* Jakarta: Central Bureau of

Statistics

[3] World Bank. (2016). *World Development Report 2016: Digital Dividends* Vol. 65. Washington: World Bank

[4] Ministry of Education and Culture. (2015). *Strategic Planning 2015-2019*. Jakarta: Ministry of Education and Culture

[5] Coordinating Minister for Maritime Affair. (2015). *Performance Report of Coordinating Minister for Maritime Affair*. Jakarta: Coordinating Minister for Maritime Affair

[6] Government of the Republic of Indonesia. (2014). *Annual Report of Central Government 2013*. Jakarta: Government of the Republic of Indonesia

[7] Government of the Republic of Indonesia. (2011). *Acceleration and Expansion of Indonesia's Economoic Development 2011-2025*. Jakarta: Government of the Republic of Indonesia.

[8] Porter, M. E. (1990). *The Competitive Advantage of Nations*. New York: Free Press

[9] Porter, M. E. (2000). *Location, Competition, and Economic Development: Local Clusters in a Global Economy*. Economic Development Quarterly. New York: Free Press

[10] Morosini, P. (2004). *Industrial clusters, knowledge integration and performance*. World Development, 32(2), 305–326.

[11] Boja, C. (2011). *Clusters Models , Factors and Characteristics*.

[12] Porter, M. E. (1998). *Clusters and the New Economics of Competition*. New York: Free Press

[13] Norman, V. D., & Venables, A. J. (2004). *Industrial clusters: Equilibrium, welfare and policy*. Economica, 71(284), 543–558.

[14] Wijnolst, N. (2006). *Dynamic European Maritime Clusters*.

[15] Wijnolst, N. (2006). *European Maritime Cluster*.

[16] Engelke, T., & Schleswig-holstein, H. L. (2009). *Regional Maritime Clusters: Brussels*.

[17] Richardson, G. P. (2011). *Reflection on the foundations of system dynamics*. Dyn. Rev.

[18] Walker, W. E., Harremoës, P., Rotmans, J., van der Sluijs, J. P., van Asselt, M. B. A., Janssen, P., & Krayer von Krauss, M. P. (2003). *Defining Uncertainty: A Conceptual Basis for Uncertainty Management in Model-Based Decision Support*. Integrated Assessment, 4(1), 5–17.

The VLSI design of the sub-band filterbank in MP3 decoding

Jia-Xin Liu and Li Luo

School of Electronic and Information Engineering, Beijing Jiaotong University Beijing, China

Corresponding author's e-mail: 15120018@bjtu.edu.cn

Abstract. The sub-band filterbank is one of the most important modules which has the largest amount of calculation in MP3 decoding. In order to save CPU resources and integrate the sub-band filterbank part into MP3 IP core, the hardware circuit of the sub-band filterbank module is designed in this paper. A fast algorithm suit for hardware implementation is proposed and achieved on FPGA development board. The results show that the sub-band filterbank function is correct in the case of using very few registers and the amount of calculation and ROM resources are reduced greatly.

1. Introduction

MP3, proposed by MPEG, is a kind of audio format with lossy compression [1]. MP3 has been widely used so far because of its high compression ratio and sound quality [2]. The sub-band synthesis filterbank is the final step of MP3 decoding, but also the most critical step. The data after IMDCT transformation can be regarded as approximately spectrum representation of the time sample values. These data will completely transform back to the time domain after the sub-band synthesis filterbank. And the original PCM samples are obtained. The calculated amount of the sub-band filterbank is very large, accounting for more than 60% of the calculation in MP3 decoding [3].

Until now, the technology of MP3 decoding is pretty mature. Currently, general structure is achieving MP3 software decoding based on DSP or RISC processor. This method is very flexible and can support a variety of audio format decoding. However, it is easier to implement but cost higher if floating-point DSP is adopted. And it is more cost-effective but harder to achieve if fixed-point DSP is utilized [4]. There are also some parallel decoding methods proposed to decrease the decoding time [5], but a lot of resources are wasted because more than one CPU is needed. Comparatively speaking, using hardware to achieve is faster and has higher efficiency and lower power consumption. In conclusion, using hardware circuit to achieve is more suitable for the sub-band filterbank module, which has a large amount of calculation. And hardware implementation can save CPU resources, hence CPU can only control other modules. Besides, the sub-band filterbank module can be integrated into IP core, and reusable in other hardware design and SOC applications. It cannot be denied that it is not flexible if hardware method is used, but compared to the advantages, achieving the sub-band filterbank in hardware method is a domain worthy of research and development.

In this paper, a fast algorithm of the sub-band filterbank is introduced, which can reduce the amount of calculation and memory room of the module effectively. And the hardware implementation of the function is given. The module is simulated in Modelsim software, and the results are compared with the decoding results using C language. And the function is implemented on FPGA development board.

2. Another section of your paper

The process of the sub-band filterbank can be divided into three parts: matrix transformation, windowing and accumulation. The first step is the matrix transformation. This process is similar to the transformation of IMDCT. Its computational formula is shown below.

$$V_i = \sum_{k=0}^{31} X_k \cos\left(\frac{\pi}{64}(16+i)(2k+1)\right) i = 0 \dots 63 \tag{1}$$

The parameter X_k represents 32 inputs taken from each sub-band, and 64 output values V_i are generated. Then the outputs are stored in a FIFO with depth of 1024. And then extract 512 sample values among them, written as U_i, the extraction formula is shown in (2) and (3).

$$U[64i + j] = V[128i + j] \tag{2}$$

$$U[64i + 32 + j] = V[128i + 96 + j] \quad i = 0 \dots 7, j = 0 \dots 31 \tag{3}$$

Then windowing transformed U_i, the formula is shown in (4). D[i] is window function, which is already given in MP3 decoding standard.

$$W[i] = U[i]D[i] \quad i = 0 \dots 511 \tag{4}$$

The last step is accumulation. The 512 W vectors after windowing can be arranged to 16 lines according to each line of 32 data. The original PCM data can be obtained by stacking together the 16 W vectors of each column. And finally, 32 PCM audio data are obtained in all. The accumulation formula is shown below.

$$PCM[j] = \sum_{i=0}^{15} W[j + 32i] \quad j = 0 \dots 31 \tag{5}$$

3. Improvement of the algorithm

The calculated amount of the sub-band filterbank is very large, and mainly concentrated in the matrix transformation part. So in order to reduce the amount of calculation, this part should be taken into consideration first. There are many fast algorithms of the matrix transformation [6-7], but most of them use a lot of multiplications and divisions. Hardware implementation isn't suitable for these algorithms because of the restrictions on accuracy of the calculation and the characteristics of the circuits and etc. Besides, the multiply-accumulator structure can also be used [8-9], but this structure will cost large chip area, a lot of hardware resources are wasted and the power consumption is very high, so it is not suitable either. So in order to simplify the matrix transformation formula, the symmetry of cosine function should be taken full advantage of. First, (1) has symmetry, the cosine function in matrix transformation is written as $C_{i,k}$, so:

$$\begin{aligned} C_{32-i,k} &= \cos\left(\frac{\pi}{64}(48-i)(2k+1)\right) \\ &= \cos\left(\left(\frac{\pi}{64}i - \frac{3\pi}{4}\right)(2k+1)\right) \\ &= -\cos\left(\left(\frac{\pi}{64}i + \frac{\pi}{4}\right)(2k+1)\right) \\ &= -\cos\left(\frac{\pi}{64}(i+16)(2k+1)\right) \\ &= -C_{i,k} \quad i = 0 \sim 15 \end{aligned} \tag{6}$$

Similarly, when i varies from 33 to 47, $C_{96-i,k} = C_{i,k}$, when i equals to 16, $C_{i,k} = 0$, when i equals to 48, $C_{i,k}$=1. Then take the cosine function back into matrix transformation formula, and we can get:

$$V_{32-i} = -V_i \quad i = 0\sim15$$
$$V_{96-i} = V_i \quad i = 33\sim47$$
$$V_i = 0 \quad i = 16$$
$$V_i = \sum_{k=0}^{31} X_k \quad i = 48 \tag{7}$$

Hence, in the calculation of a sub-band, only 32 V[i] vectors need to be calculated. The other data can be obtained directly or negatively. And the calculated amount is reduced to half. In the remaining 32 V[i] vectors, continue to observe the symmetry of the cosine function in each data that needs to be calculated. When i is fixed, we can notice that:

$$C_{i,k} = \cos\left(\frac{\pi}{64}(16+i)(2k+1)\right)$$
$$C_{i,31-k} = \cos\left(\frac{\pi}{64}(16+i)(63-2k)\right)$$
$$= \cos\left((16+i)\pi - \frac{\pi}{64}(16+i)(2k+1)\right) \tag{8}$$

So, when i is odd, we can get $C_{i,k}$ =-$C_{i,31-k}$, $i = 0\sim15$, according to the symmetry of cosine function. And when i is even, $C_{i,k} = C_{i,31-k}$, $i = 0\sim15$. So, in all 32 inputs, the kth and the 31-k[th] inputs can be extracted together. First add or subtract according to the parity of i. Then multiply with cosine function. In summary, the matrix transformation formula in the sub-band filterbank can be written as (7) and (9):

$$V_i = \sum_{k=0}^{15} (X_k + X_{31-k}) \cos\left(\frac{\pi}{64}(16+i)(2k+1)\right) \quad \text{i is even}$$
$$V_i = \sum_{k=0}^{15} (X_k - X_{31-k}) \cos\left(\frac{\pi}{64}(16+i)(2k+1)\right) \tag{9}$$

i is odd

In terms of calculated amount, if the original algorithm is used, 64*32=2048 times multiplication and 64*31=1984 times addition need to be done in the calculation of a sub-band. And if the improved algorithm is used, only 32*16=512 times multiplication and 32*(16+31)=1504 times addition need to be done. It is very convenient to implement by hardware while the calculated amount is reduced greatly. Besides, in terms of ROM resource consumption, if the original algorithm is used, 64*32=2048 cosine values should be stored. And only 32*16=512 cosine values need to be stored if the improved algorithm is used, saving 3/4 of storage space. And ROM resource consumption is reduced greatly.

4. Hardware design

4.1. Design of top module

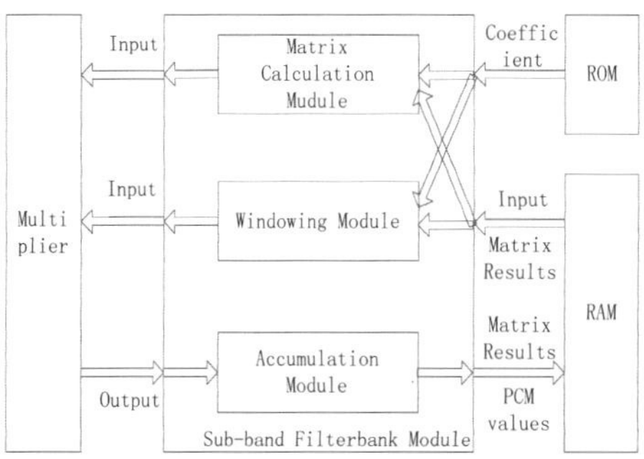

Figure 1. Design of top module of the sub-band filterbank

The design of top module of the sub-band filterbank is shown in Figure 2. In order to complete the function, a multiplier, a ROM, a RAM and a sub-band filterbank module is needed. Among them, input of the multiplier is 20-bit fixed-point number. The most significant bit is the sign bit. The remaining 19 bits represent the fractional part. Output of the multiplier is 40-bit number, the second highest bit is the sign bit and the subsequent 19-bit data represents the result of the multiplier. The simplified cosine function values and the window function values are stored in ROM. The decoding results of IMDCT are stored in RAM, for the use of matrix calculation. Meanwhile, 1024-long FIFO is also developed in RAM. The design of the sub-band filterbank module is divided into three parts. The matrix calculation module reads the data in RAM and the function values in ROM and then sends them to the multiplier. The results of the matrix calculation can be obtained by accumulating the outputs of the multiplier, and stored in the developed FIFO. The windowing module reads the results of the matrix calculation and window function values according to a specific law, and sends them to the multiplier. The PCM values are obtained after accumulating and then written to RAM.

4.2. Timing control of matrix transformation

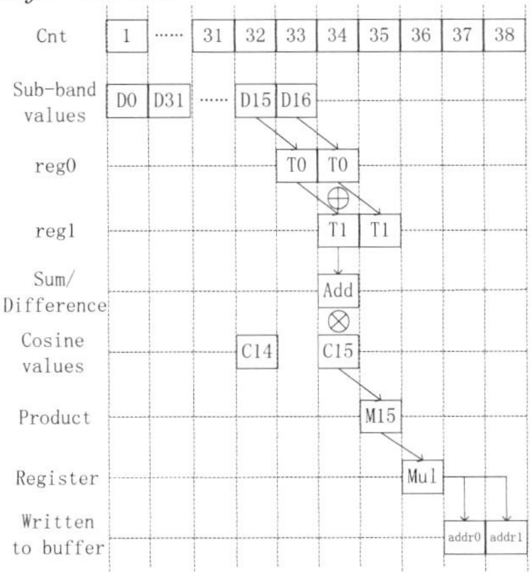

Figure 2. Timing control of matrix transformation

Timing control of matrix transformation is shown in Figure 3. When the counter value varies from 1 to 32, the addresses of kth and $31\text{-}k^{th}$ input data are given separately and the data is read out in the next clock cycle. Then in the following two clock cycles, the two values are given to two registers T0 and T1. Using combinational logic, add or subtract the two values according to the parity of i. The sum or difference is written as Add. Add multiplied with cosine function value C15 extracted from ROM and a matrix transformation result M15 is obtained. Then in the next clock cycle, the value M15 is given to register Mul. If the bit after 20 valid bits equals to 1, add 1 to its valid bit. If the bit equals to 0, just take the valid bit. In the next two clock cycles, write Mul or its negated values into two corresponding addresses respectively. At this point, the calculation of a sub-band is completed, and a granule needs 18 sets of such calculation.

The matrix transformation module owns the largest calculated amount in the entire sub-band filterbank. It can be seen from the timing control diagram that, on the premise of ensuring the correct decoding results, the symmetry of cosine function is taken full advantage of, in the same clock cycles, the amount of multiplication and addition as well as read and write operations of ROM and RAM are greatly reduced. Two matrix transformation values are decoded at one time, so the decoding speed is improved and the decoding power consumption is reduced.
Some text.

4.3. Design of FIFO

After matrix calculation, the obtained 64 outputs need to be written to a 1024-long FIFO. Then extract the i^{th}(i=0-31) of the first 32 data and the last 32 data every 128 data, multiplied with window function values and accumulate, then PCM audio samples are obtained [10]. In order to save resources and make sure that no more FIFO is called, 1024-long space is opened up in RAM, and it is used as a FIFO through controlling the addresses. The sketch map is shown in Figure 4.

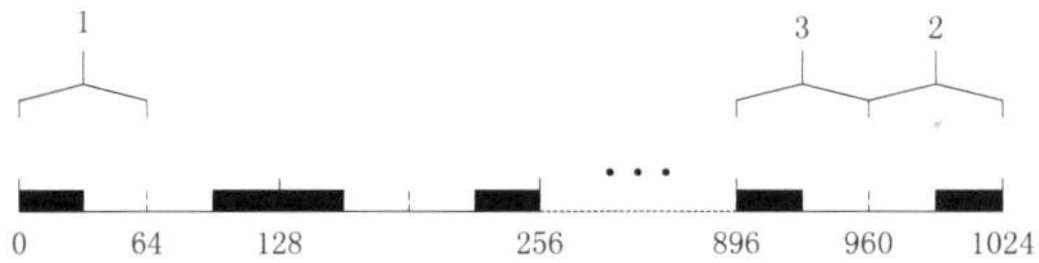

Figure 3. Design of FIFO

The first set of matrix transformation results is written to the address 1 shown in the figure above. First extract the ith data and i^{th} window function value, then extract a data every 96,32,96,32… addresses and extract a window function value every 32 addresses, multiply and accumulate 16 sets of data and the PCM values are obtained. The second set of matrix transformation results is written to the address 2. From address 960, extract data, multiply and accumulate according to the law mentioned above. All the PCM data can be obtained so.
Some text.

4.4. Simulation results and comparison

After the hardware design of the sub-band filterbank part is completed, run simulation with Modelsim. The original file used for the simulation is a MP3 file of 2 seconds. After simulation, the obtained PCM audio values are written to a new file. The PCM audio data can be played by writing a file header of wav format in the front of the new file. After verification, the playing effect of PCM audio values is consistent with the original MP3 file. In order to show the results of simulation more intuitively, the software Cool Edit can be used to view the waveform of wav files. The waveform is shown in figure below.

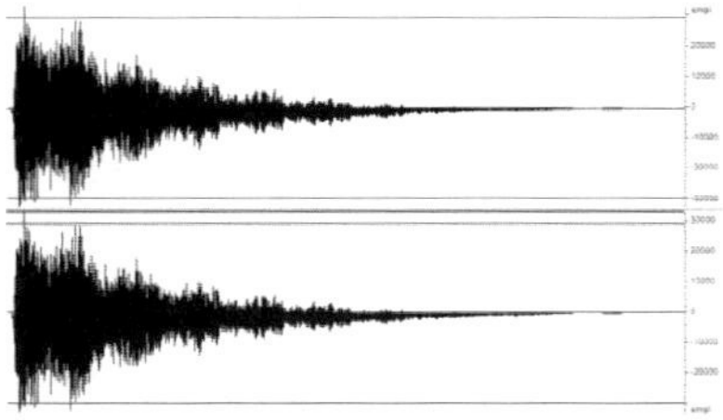

Figure 4. Waveform decoded by C language

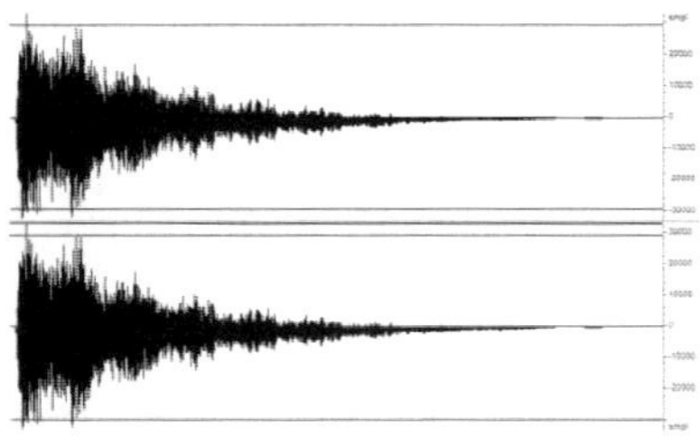

Figure 5. Waveform decoded by Verilog language

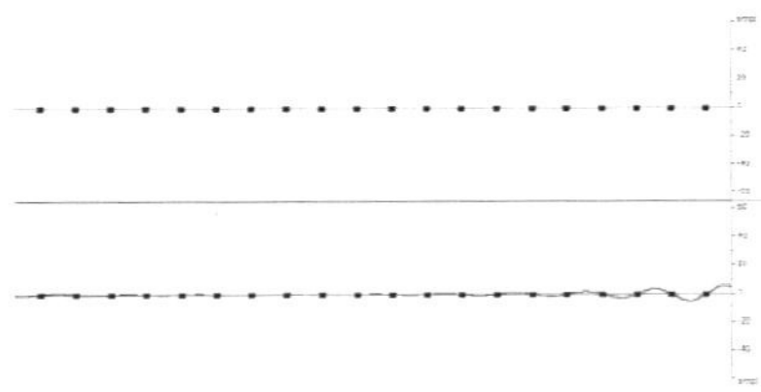

Figure 6. Waveform of the difference between two decoding results

Figure 6 shows the waveform of the results decoded by C language, the two lines are the waveforms of left and right channels. Figure 7 shows the waveforms of the two channels decoded by Verilog language. And figure 8 shows the difference between two decoding results. From the three figures above, we can see that the waveforms obtained in two different ways are basically the same, the peak is more than 30000. And the difference is very small, almost negligible. So the hardware design of the sub-band filterbank is correct.

Besides, the module is implemented on FPGA development board DE2-115 of ALTERA company. The results of analysis and synthesis and power analysis are shown in Figure 9.

Revision Name	Filterbank
Top-level Entity Name	Filterbank
Family	Cyclone IV E
Total logic elements	524
Total combinational functions	464
Dedicated logic registers	236
Total registers	236

	Fmax	Restricted Fmax	Clock Name	Note
1	173.82 MHz	173.82 MHz	Clk	

Revision Name	Filterbank
Top-level Entity Name	Filterbank
Family	Cyclone IV E
Device	EP4CE115F29C8
Power Models	Final
Total Thermal Power Dissipation	164.38 mW
Core Dynamic Thermal Power Dissipation	1.26 mW
Core Static Thermal Power Dissipation	98.60 mW
I/O Thermal Power Dissipation	64.51 mW

Figure 7. Analysis results in Quartus II

From the figure above, the sub-band filterbank module is implemented in the case of using only 524 logic units, saving a lot of resources. And the maximum clock frequency can reach 173.82MHz, completely meeting the timing sequence requirements of MP3 decoding. The input motivation of the module is the output results of the previous module IMDCT in MP3 decoding, the power consumption of the sub-band filterbank module is 164.38mW in the case of typical mode and clock frequency of 10MHz.

5. Conclusion

The algorithm of the sub-band filterbank in MP3 decoding is studied in this paper, and it is simplified by taking full advantage of the symmetry of cosine function. And the hardware design of the sub-band filterbank is given according to the characteristics of hardware circuit. The calculated amount and memory resources are reduced greatly on the premise of ensuring the decoding results are correct. Besides, the implementation of the function on development board DE2-115 of ALTERA company is given in this paper. The results show that, the function is realized correctly in the case of using very few registers.

6. Acknowledgment

This work was supported by National Natural Science Foundation of China (U1431119).

References

[1] S. Shlien, "Guide to MPEG-1 audio standard," in *IEEE Transactions on Broadcasting*, vol. 40, no. 4, pp. 206-218, Dec 1994.

[2] H. G. Musmann, "Genesis of the MP3 audio coding standard," in *IEEE Transactions on Consumer Electronics*, vol. 52, no. 3, pp. 1043-1049, Aug. 2006.
[3] B. P. Gangamamba, N. S. Murthy and P. Muralidhar, "Low power reconfigurable sub -band filter bank ASIC for MP3 decoder," 2008 International Conference on Electronic Design, Penang, 2008, pp. 1-5.
[4] Jing Chen and Heng-Ming Tai, "Real-time implementation of the MPEG-2 audio codec on a DSP," in *IEEE Transactions on Consumer Electronics*, vol. 44, no. 3, pp. 866-871, Aug 1998.
[5] W. Xue, S. Sun and X. Li, "Overlapping MP3 Multi-thread Parallel Decoding," *2015 3rd International Conference on Applied Computing and Information Technology/2nd International Conference on Computational Science and Intelligence*, Okayama, 2015, pp. 458-463.
[6] Byeong Gi Lee, “A new algorithm to compute the discrete cosine transform”, *IEEE Transactions on acoustics, speech and signal processing*, 32(6), 1984.
[7] M. D. Valdes, M. J. Moure, J. Dieguez and S. Antelo, "Hardware solution of a polyphase filter bank for MP3 audio processing," 2008 IEEE International Symposium on Industrial Electronics, Cambridge, 2008, pp. 1225-1229.
[8] T. Geng, L. Waeijen, M. Peemen, H. Corporaal and Y. He, "MacSim: A MAC-Enabled High-Performance Low-Power SIMD Architecture," *2016 Euromicro Conference on Digital System Design (DSD)*, Limassol, 2016, pp. 160-167.
[9] D. De Caro, N. Petra, A. G. M. Strollo, F. Tessitore and E. Napoli, "Fixed-Width Multipliers and Multipliers-Accumulators With Min-Max Approximation Error," in *IEEE Transactions on Circuits and Systems I: Regular Papers*, vol. 60, no. 9, pp. 2375-2388, Sept. 2013.
[10] Y. Jhung and S. Park, "Architecture of dual mode audio filter for AC-3 and MPEG," in *IEEE Transactions on Consumer Electronics,* vol. 43, no. 3, pp. 575-585, Aug 1997.

IOP Conf. Series: Earth and Environmental Science **127** (2018) 012021 doi:10.1088/1755-1315/127/1/012021

Development of intelligent monitoring purifier for indoor PM 2.5

Guanting Lou, Rong Zhu, Jiangwei Guo and Yongqing Wei

School of Instrument Science and Opto-electronic Engineering, Hefei University of Technology, Hefei, Anhui 230009, P. R. China

Corresponding author's e-mail: louguanting@gmail.com

Abstract. The particulate matter 2.5 (PM2.5) refers to tiny particles or droplets in the air that are two and one half microns or less in width. PM2.5 is an air pollutant that is a concern for people's health when levels in air are high. The intelligent monitoring purifier was developed to detect indoor PM2.5 concentration before and after purification and the monitoring data could be displayed on the LCD screen, displaying different color patterns according to the concentrations. Through the Bluetooth transport module, real-time values could also display on the mobile phone and voice broadcast PM2.5 concentration level in the air. When PM2.5 concentration is higher than the setting threshold, the convection fan rotation and the speed can be remote controlled with mobile phone through the Bluetooth transport. Therefore, the efficiency and scope of the purification could be enhanced and further better air quality could be achieved.

1. Introduction

PM 2.5 in the air reduces visibility and causes the air to appear hazy when its levels are elevated. PM 2.5 pollution has raised serious concerns for public health. Currently, the outdoor PM2.5 monitoring technology has been developed more mature, and outdoor individual protection could be achieved by facial masks. In the daily mobile phone weather forecast we can observe the monitoring value of PM2.5 in the air. Indoor air purification usually relies on expensive and energy-intensive air-filtering devices and a wide range of purification possibilities and improves of the air situation need long-term plan to achieve as well [1-3]. PM2.5 pollution is particularly harmful since it can penetrate human bronchi and lungs owing to the small particle size. There have been serious PM pollution problems in developing countries, such as China. Students or office workers often stay indoors every day for more than 13 hours, the indoor air treatment/purification would not only improve the air quality in the living environment, but also solve the issues caused by PM2.5 disease incidence.

However, the price of indoor PM2.5 air purifier currently sold in the market is still high, and their functions cannot fully meet the customer requirements and need to be enhanced. For the market dominated by students and office workers, low-cost high efficiency PM2.5 air purifiers are demanded and it needs to be ensured that the function should be as complete as possible to avoid duplication consumption. This investigation was combined with the level of consumption and living conditions of students to develop a small simple PM2.5 monitoring purifier, which integrates the functions of monitoring, air purification, voice broadcast, Bluetooth data transportation, and air convection. The objective is to improve the air quality in student dormitory and ensure the comfort of college students living environment.

IOP Conf. Series: Earth and Environmental Science **127** (2018) 012021 doi:10.1088/1755-1315/127/1/012021

2. Indoor PM2.5 data monitoring and determination of air quality

Based on Arduino single chip, equipped with dust sensor (GP2Y1010AU0F SHARP) including conversion the parameters generated by the relevant devices, PM2.5 concentration monitoring is achieved. According to the national PM2.5 concentration segment and the corresponding air quality index, the concentration level grading is categorized through the design procedure and it is convenient to view the monitoring value of the PM2.5 concentration from the intelligent monitoring purifier for indoor PM2.5.

The PM2.5 detection principle is schematically described in Figure 1. The sensor center hole allows the sufficient free air flow. The directional launched LED light is using as source for detection of dust content in the air, which is determined via the light refraction after the light penetrates though the dust. The 24-hour mean value of PM10 in the air was obtained with the aid of the dust detection system.

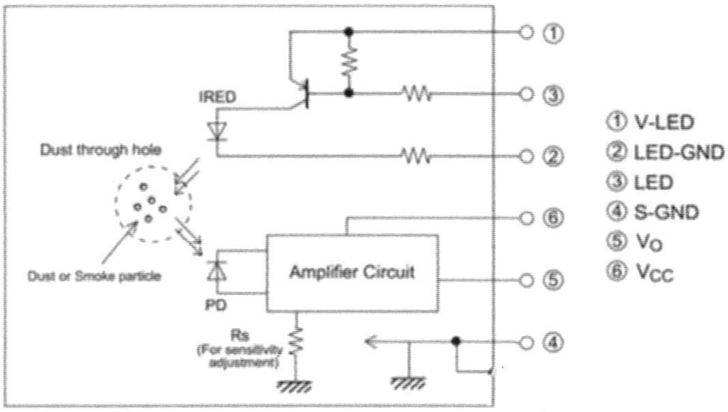

Figure 1. The diagram of the PM2.5 detection principle for the dust sensor (GP2Y1010AU0F SHARP).

The value corresponding to the dust sensor voltage is converted into dust density while the PM10 dust density into the PM2.5 dust density, the real-time data output to the monitoring device. The air quality level [4] is translated according to the 24-hour average of PM2.5 as shown in Table 1.

Table 1. The index of indoor air quality, category and color codes of air quality index

Index of Air Quality Level (IAQL)	PM2.5 24-hour Mean ($\mu g/m^3$)	Index of Air Quality (IAQ)	Level of IAQ	Category and Color Codes of Air Quality Index	
0	0	0~50	1st class	Excellent	Green
50	35				
100	75	50~100	2nd class	Good	Yellow
150	115	101~150	3rd class	Light Polluted	Orange
200	150	151~200	4th class	Medium Polluted	Red
300	250	201~300	5th class	Heavy Polluted	Purple
400	350	>300	6th class	Server Polluted	Dark Red
500	500				

In our investigation, Sharp dust sensor is used due to its cheaper price and easy use. Therefore, the entire product cost of the air purifier for indoor PM2.5 can be rational and acceptable easily by customers. Meanwhile, the program rewrite through the conveniently accessed process of parameter setting/resetting greatly not only enhance the writing ability of software design, but also improve the reliability of the molding products.

The software modules included Chinese characters are implemented to complete the programming of the liquid crystal display; the continuous improvement can be completed in the subsequent debugging and continuous testing. Ultimately, a complete dust sensor is applied to monitor the value and achieve the PM2.5 numerical conversion. Meanwhile, a display pattern is obtained based on the level of PM2.5 in the indoor air, the liquid crystal display can be based on PM2.5 numerical segment shows different expressions. The matrix diagram of the liquid crystal display principle is shown in Figure 2.

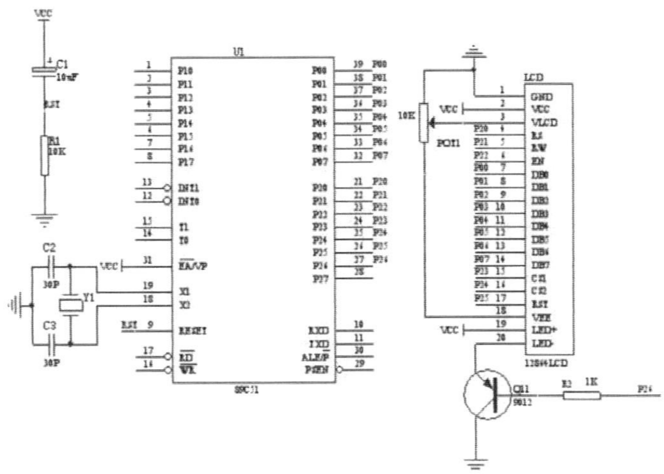

Figure 2. The matrix diagram of the liquid crystal display principle

3. Voice broadcast of PM2.5 concentration level

SYN6288 TTS serial speech integration module is used to support a variety of text code formats including GB2312, GBK, BIG5 and UNICODE. Therefore, the device can generate any Chinese text and support English alphabet text also. With intelligent text analysis and processing algorithms of those modules, the voice broadcast component can correctly identify the value, number, time and date and commonly used weights and measures units. It supports a variety of text control tags to enhance the correct rate of text processing. It also supports sleep function mode, in the sleep state the device can switch to another operation to reduce power consumption. The device supports a variety of ways to query the working state of chip and supports 16 volume adjust. The volume of the playback text and the volume of the background music can be controlled separately. The word speed of playback text can be adjusted by sending the control mark with 6 speed adjustment. The module pin structure and integrated connection of the microcontroller is show in Figure 3.

IOP Conf. Series: Earth and Environmental Science **127** (2018) 012021 doi:10.1088/1755-1315/127/1/012021

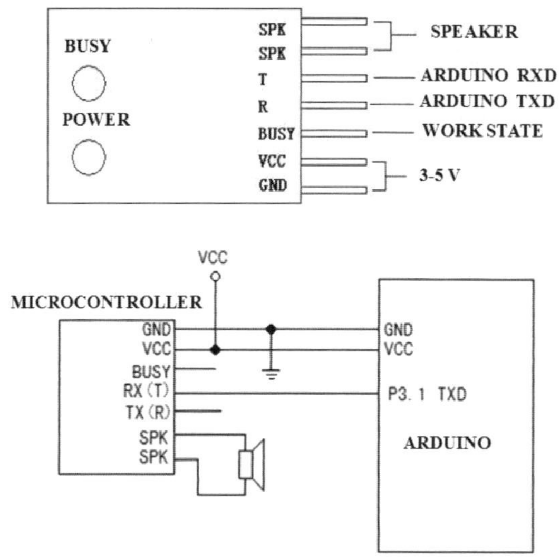

Figure 3. The matrix diagram of the liquid crystal display principle

Through SYN6288TTS serial speech integration module, the dust sensor output data can be successfully converted to voice and thus the real-time PM2.5 monitoring is achieved with the intelligent output. The main difficulty encountered in this work is time jump of PM2.5 monitoring data and the evolution value is not a fix number. Before and after the broadcast, the value of the PM2.5 concentration may change, sometime vary greatly, most likely even not in a same air quality level. Therefore, the broadcast method was adjusted from the instantaneous voice broadcast value to delayed broadcast value in order to avoid the time jump value caused inaccuracy and the air quality level is added. According to the value of air quality improvement or a slight reduction and additional the level broadcast, the owner has a general grasp of the indoor air quality. The final output is that the voice broadcast is at appropriate speed and easy to listen with the well-designed 16-level volume and 6-voice speed adjustment.

4. Indoor air purification for PM2.5

Multiple function-layer filters followed after the turbine fan is designed to guarantee the best indoor air purification as shown in Figure 4. The first layer of the purifier, non-woven air filter cotton, captures larger particles in the air, such as dust and hair; The second layer of activated carbon acts as the adsorbate for odor and harmful gases; The third layer of HEPA filter can capture the PM as small as 0.3 microns (PM = 0.3).

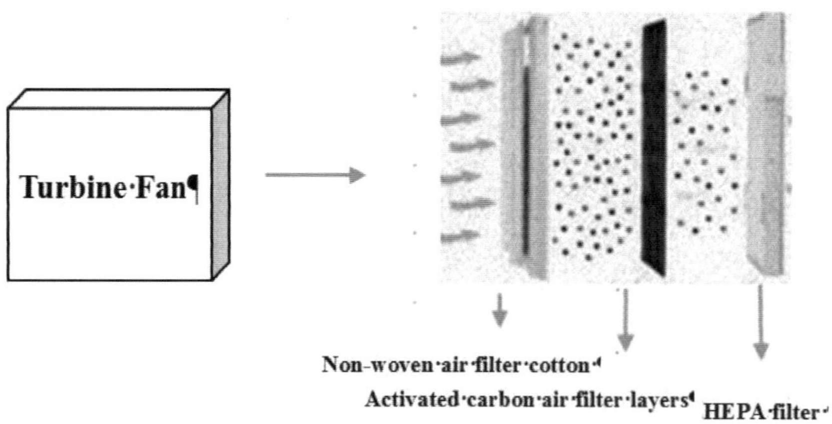

Figure 4. The schematic diagram of the multiple function-layer filters of the air purifier.

Turbine fans provide the sufficient air flow through the air purifier. With higher turbine fan speed, air flow volume increases and thus the purification for PM2.5 speeds up. Furthermore, the efficiency of purification for PM2.5 improves also. The initial filter and the activated carbon filter pretreat the air through turbine fan; they also reduce the damage for the subsequent HEPA filter caused by bacteria and reduce the unnecessary loss. At the same time, we sealed gap between the filter and the device box to ensure the accuracy of the collected data after purification. That makes sure that unpurified air cannot penetrate the filter to disturb PM2.5 numerical monitoring. This kind of design/set-up keeps the reliability of monitoring data for the subsequent analysis and calculation with great convenience.

According to the dynamic experimental results shown in Table 2, the device can achieve the air purification efficiently. To avoid damage to the purification device, the device should be kept far away from smoke environment. The timely replacement of the early filter and activated carbon filter are necessary to maintain the efficiency the indoor air purification.

Table 2. The dynamic testing results of air purification for indoor PM2.5

Experimental Results* Concentration / Time	1st	2nd	3rd
Before Purification	289	276	293
30s	246	226	249
50s	237	209	227
100s	192	183	170
200s	96	92	89
300s	45	43	39
Efficiency (%)	84.4	84.4	86.7

*Note: The simulated testing environment is 500x500 mm^2

5. Bluetooth data transformation between mobile phone and data transfer

HC-05 Bluetooth serial module is used to connect the Arduino single chip computer [5, 7]. Thus, communication with Android system, real-time monitoring and the data of long-distance transportation can be realized. Because Arduino has only a pair of clear serial port for voice module communication, we access the information through the Arduino with Software Serial Library function and set the 4,5 port as a soft serial port to achieve real-time communication. Through the mobile app, the real-time PM2.5 value has been consistently displayed. According to the program, the voice module can broadcast the corresponding real-time air quality level simultaneously as well.

6. Convection fan design and intelligent fan speed control

For further better indoor air purification results, with two small fans placed on the purifier and in the opposite direction, we set-up convection fan system to enhance air circulation. When the indoor PM2.5 value is high, the convection fans are turned on and air convection increase. Thus, the efficiency and scope of purification for indoor PM2.5 can be greatly enhanced, and the surrounding air quality is improved also.

HC-05 Bluetooth module equipped with fan is still used for state transmission and semi-automatic control. Exhaust fan is added for intelligent speed control to coordinate the efficiency and balance the noise. When the air quality is good, the purifier will automatically set the speed to 0 for energy saving. The PWM frequency for the coming Arduino is only 256, and this does not meet the fan speed at 20KHZ. We modified the timer pre-bit and PWM output mode, so that the frequency of 3IO just meets the requirements.

7. Conclusions

We have developed the intelligent monitoring purifier with high removal capacity and efficiency over 84% for indoor PM2.5. Integrated with the removal of PM2.5 and even smaller particles with the aid of HEPA filter and the advanced multi-functions achieved by electrical devices, including the intelligent monitoring, voice broadcast for the PM2.5 concentration, and Bluetooth data transportation.

It is believed that indoor PM2.5 air purifier with intelligent monitoring system can be used as a stand-alone device to achieve a healthier indoor living environment.

8. Acknowledgement

The project is supported in part by the Undergraduate Innovative Discipline Project in China (2015CXCY337).

9. References

[1] Liu C, Hsu P, Lee H, Ye M, Zheng G, Liu N, Li W and Cui Y,2015 Transparent air filter for high-efficiency PM2.5 capture, *Nature Communication,* 6(6205), pp 1-6

[2] Chen C, Li X and Yan J, 2014 Study on indoor pollutant removal approaches for building in Tianjing Eco-city, *Environment, Energy and Sustainable Development* (Taylor and Francis Group), 31. Sung W, Kao J C M, Chen R (Eds)

[3] Robles R J, and Kim T, 2010 Applications, Systems and Methods in Smart Home Technology: A Review, *International Journal of Advanced Science and Technology*, 15, pp 38-47

[4] Chen W, Wang F, Xiao G, Wu K and Zhang S, 2015 Air quality of Beijing and impacts of the new ambient air quality standard, *Atmosphere*, 6, pp 1243-1258

[5] Wu Y, Li L, Chen S and Pen W, 2014 Design of photoelectric encoder detector based on Arduino development environment, *Modern Electronic Technique,* 37(2), pp. 124-125

[6] Cui Y, Zhang W and Bai Y, 2013 An intelligent home control system based on Arduino, *Application of Electronic Technology*, 39(9), pp. 54-60

[7] Lian K and Hsiao S, 2013 Intelligent multi-sensor control system based on innovative technology integration via ZigBee and Wi-Fi networks, *Journal of Network and Computer Applications,* 36(2), pp 756-767